What in the World Are PET Scans?

What in the World Are PET Scans?

Written By:
Austin Mardon
Razan Ahmed
Katerina Bavaro
Kendall Caperchione
Jessica Henschel
Hassan Khan
Ruchira Nandasiri
Megha Sharma
Rosaline Sullivan
David Supina
Kelly Wu

Editor:
Stephanie Lazar

Cover Design By:
Josh Harnack

First Printing: 2021

Typeset and Cover Design by Josh Harnack

ISBN: 978-1-77369-232-6

Golden Meteorite Press
103 11919 82 St NW
Edmonton, AB T5B 2W3
www.goldenmeteoritepress.com

Table of Contents

What Was Medical Imaging Like Before PET Scans?
By Kelly Wu

Introduction

A man lies still in the tube-like machine, sounds of clicking and buzzing filling the space. In the dark room on the opposite side of the glass is a model of the patient outlined across the monitors, colour blossoming across his chest. A bright glow highlights the worst of it, collecting on a particular blinding spot near the lung that informs the panel of doctors that their patient is cancer ridden and requires treatment. Luckily, they have caught it in its early stages and this procedure has saved the patient from what would have been described a hundred years ago as a certain death sentence. Medical imaging refers to the different technologies and processes used to view the interior anatomy for the purposes of identification, controlling, and treatment of medical conditions. Radiography in specific, is the field of medical imaging that involves high energy radiation. While this patient had a relatively comfortable PET scan experience, in which he lays still in the machine following an injection, traditional diagnosis methods were not always this clinical. Whereas modern technology has offered patients much more of a guarantee of safety, the same could not be said even for merely a few decades in the past (Freiherr, 2014). Some recorded techniques even caused more harm than good. Nonetheless, each prior invention, regardless of the risk, was a step forward in perfecting the accuracy and security of medical imaging to the pinnacles of technology we have today.

Pre-1800 & History of Anatomical Dissection

Historically, much of our medical knowledge came by dissection. With the first recorded scientific human dissection being conducted as early as the third century B.C., this makes anatomy as the oldest scientific medicinal discipline ("The History of Anatomy - from the beginnings to the 20th century", n.d.). Leonardo Da Vinci and Michaelangelo, the great Renaissance painters themselves, also had a hand in guiding the scalpel. They were not content to sit idly by and observe postmortem studies conducted by their trained friends, but rather played an active role in order to become better artists. Da Vinci in particular examined around thirty cadavers, with some even believed to have been stolen from cemeteries, resulting in accompanying drawings that were anatomically accurate (Riley, 2019). Entering into the 17th century, entire theaters were built devoted to dissection that were open to the public who were eager and curious about the inner workings of the body. While this practice of careful dismemberment was especially beneficial to our understanding of the structure and functions of the body, it was less effective in diagnosis, prevention, and treatment. As such, the concept of imaging intrigued doctors and scholars alike as it represented a solution that avoids death as a means to better study a disease.

Wilhelm Roentgen and Invention of the X-Ray (1895)

This ambition first emerged as a reality in 1895 with Wilhelm Roentgen's invention of the X-ray (Chodos & Ouellette, 2001). Roentgen was a German mechanical engineer who discovered that ionizing radiation was able to penetrate most materials, save for bones and metal. By placing a photosensitive plate behind the subject of the radiation, a skeletal image could be captured. The first medical radiograph was of Roentgen's wife and the film can be seen to this day, the dark print of her fingers and wedding ring a stark contrast against the yellowing background. In just a few months after his breakthrough, X-rays were first used clinically in the United States when Dr. Edwin Brant Frost created an imaging of a Colles fracture ("First Clinical X-ray in America Performed", n.d.). Culturally, the apparatus was of interest to those beyond the physics and medicinal community. Shops and studios opened in the business of taking "bone portraits" and references to X-rays often found themselves being mentioned across journals, short stories, political cartoons, and advertisements alike. While those fads have long faded away, the scientific impact has lingered on ever since. Given the different densities of bodily tissues, Roentgen's experiments with this newly discovered light phenomenon revealed abnormalities and allowed early detection of conditions when analyzing the developed plate. His refusal to take out any patents on his device gave the public free use of his work, ultimately laying the foundations for further progress in the field of medical imaging.

Despite its ingenuity, Roentgen's innovation was not without its flaws. Both the technician and patient were exposed to extremely high levels of radiation throughout the long duration it took to capture the image. Compared to modern day X-rays where lead coverings are provided to prevent unnecessary exposure and have a duration of less than a second, the 1896 version of the process took nearly 90 minutes and subjected all parties within range to nearly 1,500 times the amount of radiation (Hutchison, 2011). This led to symptoms such as swelling, burns, blisters, loss of hair, and in the worst and most ironic cases, cancer (Sansare, Khanna & Karjodkar, 2011). Unfortunately, doctors and researchers who oversaw this procedure often fell victim to early death themselves.

Improvement in Medical Imaging in the Early 20th Century

As the medicinal community progressed into the 20th century, attempts were made to improve the X-ray with a forgotten household item we may not think twice of today — film cassettes. What is now often remembered as a fun trinket from the 80s was once a part of the refinement process which reduced the time of exposure. Similar to film photography, a roll of film was loaded into the cassette in a darkroom. The patient stood unmoving for up to eleven minutes while the radiation was projected, resulting in a developed image on the film cassettes (Goel & Bell, n.d.). Further down the line in 1946, doctors believed they could minimize patient exposure with intensifying screens (Haus & Cullinan, 1989). The film used to capture the image was treated with fluorescence such that it would become more sensitive to the radiation, producing images with high definition. In this way, technicians could further lessen development time by wearing special glasses to preview the fluorescent radiograph ("Intensifying screens. Problems and developments", 1955). However, this solution did not last long as it only transferred the exposure time that patients avoided onto the doctors who were directly viewing the radiation images themselves.

Two discoveries that helped the radiology safety movement gain momentum were the discoveries of the television and contrast medium. Television and moving images allowed doctors and technicians to direct the X-ray process at a distance, minimizing radiation exposure. The X-ray image intensifier (abbreviated as XRII) converted X-rays at a higher intensity compared to the traditional fluorescent screens used in the 40s. Following the rise of technology, it was then converted to be compatible with the use of television monitors ("The Image Intensifier (II) | Radiology | SUNY Upstate Medical University", n.d.). Contrast mediums on the other hand enhance and distinguish the interior anatomy when scanned by the X-ray ("Contrast Media", 2016). The contrast agent used today was first introduced in the 1950s and was designed to be ingested orally.

Used in tandem, these two inventions propel the field of radiology into a new era, with widespread use of nuclear medicine for diagnostic means beginning in the 1950. Mirroring the steps of a present day scan, the patient intakes a combination of radionuclides and pharmaceutics that act as tracers (Drozdovitch et al., 2014). The chemical mixture enters the bloodstream, gathering in parts of the body and cell groups which are particularly active and may be a cause for concern. The image is recorded by a gamma camera with the potential problem zones showing up as highlighted areas, allowing for early detection for medical conditions (Advancing nuclear medicine through innovation, 2007). Presently, nuclear medicine is predicted to become a $12.8 billion dollar industry, one that spans across multiple concentrations including cardiology, psychiatry, neurology, and oncology ("Nuclear Medicine Market Value Projected To Reach US$ 12.8 Billion By 2027: Acumen Research And Consulting", 2021).

Karl Dussik and the Invention of Sonography (1956)

With the conclusion of World War Two, the extensive use of the sonar as a method to detect enemy ships slowly morphed into a technique intended for more life affirming uses. Unlike X-rays, a sonor operates by using high frequency sound waves as opposed to ionizing radiation ("The History of Ultrasound | BMUS", n.d.). Radiologists would place a transducer on the patient's area of concern and rather than using film or a screen, high frequency sound waves would be launched through a probe towards the body, which would then be reflected back to the source. An image emerges on a monitor after the reflected waves are converted into electrical signals. The first recorded medical use of the sonar was in 1947 and is credited to Karl Dussik, a pioneer in neurology. Karl and his brother Friedrich constructed the first early contraption that fitted the patient's head snug in between two transducers. Directing the ultrasound beams towards the skull, he attempted to visualize the interiors of the brain for the purpose of identifying tumors (Woo, 2015). In 1956, nearly a decade later, obstetrician Ian Donald and engineer Tom Brown introduced sonar use to what we now contemporary know as the ultrasound. Turning the frequency towards capturing the image of fetuses and a tumor along the female genitalia, their work allowed technicians to visualize tissue density (Nadrljanski & Bell, 2021). The low risk and non-invasive procedure launched into global use by the end of the 20th century, finding itself particularly useful in the areas of obstetrics and gynaecology. Used to image a fetus within the womb, it can verify pregnancy, sex, and identify fetal irregularities. Socially, it is difficult to imagine a modern pregnancy without undergoing some form of obstetric ultrasonography.

Godfrey Hounsfield and the Invention of the CT Scanner (1972)

Following Dussik's innovation, much of the imaging devices we are familiar with today began to pick up traction from the 60s and onwards. The CT scanner, also referred to as computerized tomography or computerized axial tomography (CAT), was first released commercially in 1972 by Godfrey Hounsfield, a British engineer ("History of the CT Scan | Catalina Imaging", 2021). Founded on the mathematical theories laid by Johann Radon and Stefan Kacxmarz, Hounsfield's apparatus captured an object in a series of images in slices, such that the object could be reconstructed visually. Utilising ionizing radiation, a computer, and contrast materials, an image is produced that is of increased detail compared to a regular X-ray alone. Interestingly, it has been widely suspected that the owners of Abbey Road Studios, the company behind the renowned band The Beatles, were partly responsible for the creation of the CT scanner. It has been rumored that they used a portion of the overwhelming album sales in the 60s to fund foundational research. The first medicinal use of the scanner was recorded to be Dr. James Ambrose in 1971, who was treating a woman believed to have a brain tumor. While a CT scan can range from 10 to 30 minutes presently, it took Ambrose numerous days to complete as the scanner took hours to reconstruct just a single image slice from the raw data. After the initial scan, it then further took another several days to develop the overall radiographic compilation of the patient (Ambrose, Gould & Uttley, 2006). From then on, the discovery of the CT scan became an invaluable addition to the medicinal imaging community, with a recorded use of 3 million scan examinations by the 1980's ("Half A Century In CT: How Computed Tomography Has Evolved — ISCT", 2016). Yet opposed to the ultrasound, CT scans are slightly higher in risk as they utilize X-rays and are not suitable for children or pregnant individuals.

Damadian, Lauterbur, Mansfield - Fathers of the MRI

As are most inventions, it is difficult to credit the idea to a single creator. Such is in this case, where the individual research of Dr. Raymond Damadian, Paul Lauterbur, and Sir Peter Mansfield collectively contributed to the achievement of magnetic resonance imaging. Lauterbur conducted experiments visualizing differences between cancerous and non-cancerous cells with magnetic resonance on dead tissue, publishing his theory in 1971 and producing the first MR image. Mansfield also discovered similar findings that would allow a 3-D reproduction of the image. In the end, Dr. Raymond Damadian won the race to construct the first whole body human scanner, the prototype of what we now know as the MRI in 1977("When and why was MRI invented", 2019). Building on the phenomena that when placed in a magnetic field, the nucleus of atoms will vibrate in response to the application of a secondary oscillating magnetic, Damadian built on Lauterbur's applications. Given that cancerous cells retain more water and would then contain higher amounts of hydrogen atoms, the magnetic resonance would be able visualize a difference among normal tissues and tumors. Similar to an ultrasound, it avoids ionizing radiation and contrast material, bypassing the biggest risks associated with X-rays. However, all metal objects must be removed prior to entering the MRI lest the patient be subjected to the projectile effect — that is, the pulling of the magnetic object into the core of the MR machine (Campbell, 2015). With the aid of computers, the MRI was able to produce higher quality images and increased detail on finer structures, all under a short period of developing time. Damadian's machine produced photographs that were of increased detail in comparison to X-ray and CAT scanner versions, ushering the radiology community into a new age of medical imaging.

Conclusion

Though not entirely primitive, the procedures listed above pale in comparison to the PET scan. They are not only less sensitive to detecting metabolic changes on a cellular level within the body, but are presented as 2-D images, whereas PET scans present a 3-D model ("PET Scan: Tests, Types, Procedure", 2021). Nonetheless, from tweaking radiation leadened X-rays to deriving alternative uses from devices of art and war, medical imaging has come a long way. Paved by curious researchers, unsuspecting patients, and doctors with good intentions, the path taken to reach the clean, clinical version of radiography we now know today is paid by their blood and sacrifice. While the process of understanding the hidden secrets of our anatomy has been refined over the past century such that radiation doses have been lowered to acceptable amounts, photographs are in significantly sharper detail, and improvement in overall technology has allowed for easy access and construct of these images, there is much to be said about the countless victims and tireless scholars that have lost their lives for us to get to where we are presently. So as we begin to delve into the ins and outs of the positron emission tomography scan, we implore you to keep these unnamed and unsung heros who have inadvertently saved millions of lives down the line.

Who invented or discovered PET scans?

By Kendall Caperchione

Introduction

As defined in the following chapter, Positron Emission Tomography (PET) is a type of physical technology that uses medical imaging to detect and reveal how various muscles, tissues, and organs are functioning in the body (Mayo Clinic, 2020). Despite this instrument only recently emerging within the last fifty to sixty years, medical tomography and internal imaging has become one of the most mainstream methods for physicians when observing patients for disease.

The end of the Second World War introduced nuclear technology in different fields, including medical nuclear advancements that matured imaging and disease detection. Medical imaging with gamma waves was something new to the industry, and with the need to advance nuclear technology after the war, emission tomography became attractive to scientists and developers alike. Many scientists and doctors can be credited for the development and progression of tomographic imaging, and David E. Kuhl is at the top of that list. Arguably considered the modern 'grandfather' of tomographic imaging, Kuhl completed a successful application of rotational tomography in the 1950's (Wackers, 2019).

The 1950's also saw the first successful demonstration of annihilation radiation at the Massachusetts General Hospital. Gordon Brownell and his team proposed through nuclear positron emission that they could enhance the quality of brain imaging, particularly resolution of the sections of the brain (CERN, 2021).

This demonstration would further become the foundational experiment for the science behind positron emission tomography, as the Brownell group was able to produce the 'how' regarding the conduction of the scan (see chapter 6).

Over the following two decades positron tomography would continuously be developed and tested in medical imaging. Michael Ter-Poggossian and colleagues, based out of Washington University's School of Medicine, saw the progress in nuclear science and tomographic imaging. Ter-Pogossian and his colleagues in unison with Brownell began constructing what would become the structural basis of the modern PET scan machine (Portnow, Vaillancourt & Okun, 2013). Their design was influenced by Kuhl's desigans in PET I-III, and with the introduction of ringed PET machine by Ter-Pogossian and his colleagues in 1976, positron emission tomography was ready to be tested and used in the medical field as the primary method of nuclear imaging.

1976 also marked the first human PET scan performed by a team of scientists including Abass Alavi, a physician-scientist who under the instruction of Kuhl and other professors was the first to administer a radiotracer to a patient in hopes of conducting imaging of the brain and body (Alavi, 1979). Alavi would later use PET scanning to detect different forms of diseases and cancers using radiotracers, and would continue researching through his career with the University of Pennsylvania, eventually becoming a fellow and Chief of the Division of Nuclear Medicine (University of Pennsylvania, 2004).

The development of positron emission tomography scanning and imaging technology in the medical field has provided scientists and doctors with the necessary tools to detect and proactively respond to life threatening diseases and illnesses. Every person involved in the process of planning, developing, and creating positron emission tomography scans will be credited in history with one of the most efficient and effective ways of disease detection and medical imaging, and the person it can be traced to first is Dr. Kuhl.

David E. Kuhl and the Beginning of Tomographic Imaging

Dr. David E. Kuhl (1929-2017), born in the fall of 1929 in St. Louis, MO grew up in Pennsylvania with his father, who was an engineer at a car company during the depression. Kuhl knew from a young age that he wanted to follow a path that varied from his father, and decided on science and medicine, as he was extremely interested with physical and biological chemistry through his primary and secondary education (Wacker, 2019). When completing his post-secondary education, Kuhl studied at Temple University in Philadelphia hoping to obtain a Bachelor of Science. He then later went to the University of Pennsylvania for his medical doctorate, completing his internship, residency, and a fellowship in the department of radiology (Dunnick, 2017). Kuhl's desire to work with ratio pharmaceuticals (specifically nuclear isotopes) began when his residency led to his investigations with rectilinear scanners. Kuhl found in his research that rectilinear scanners were a danger to living patients since the device created an unsafe amount of radioactivity within the patient when the scan was conducted (Wacker, 2019). The rectilinear scanner, similar in theory to the later PET scan, was a device used for imaging a patients body and head, using radioactive material on a scan bar to detect abnormalities and reactions (Patton, 1980). The rectilinear scanner was also not desirable to Kuhl as the machine was not as effective in imaging resolution compared to the X-ray device. This provided Kuhl with an opportunity to break through in the nuclear medicine field as the work yet to be done on radionuclide scans was promising (Patton, 1980).

Kuhl wanted to research and develop other methods that could be used as an alternative to the rectilinear scanner, and began using previous X-ray testing done in the 1930's to complement his idea. Kuhl began working with principles and theories from the X-ray scanners and attempted to apply them to emission imaging using radiotracers or radionuclides (Wacker, 2019). During his residency, Kuhl developed a photo scanner device, which used radioisotope emission to produce gray scale images that were more clear and resolute than what had previously been seen in scanning machines that predated PET (Portnow, Vaillancourt & Okun, 2013). Kuhl worked alongside Roy Edwards and developed the first scanner in 1959 titled 'Mark II' which was used for emission tomography later in 1963 (Webb, 1990). The 'Mark II' scanner was possibly one of the first ever CT (computer tomography) scans ever as it used an "optical integrator" to image the patient from multiple angles using multiple scanning bars that acted as a transaxial section tomographic imaging machine (Webb, 1990). Kuhl and Edwards continued to work together to further evolve and produced their 'Mark II' machine, and were able to produce 'Mark III' and 'Mark IV' in the 1970's which would become the models for the future SPECT, PET, and CT scanning machines (Wacker, 2019).

As a practicing doctor, Kuhl further wanted to research and focus his work in the field of tomographic imaging, and did so through collaborations with other scientists and doctors abroad. Most notably, Kuhl was able to work with Dr. Louis Sokoloff from Bethesda and Dr. Alfred Wolf from the Brookhaven National Institution. Their time together created an environment for innovation, providing the first F18 (fluorodeoxyglucose)(FDG) radio tracer in a tomographic brain scan at the University of Pennsylvania (Wacker, 2019). Kuhl's development with Wolf and Sokoloff was unprecedented, as it forced the medical field into an era that used isotopes for medical imaging where the atoms, when combined properly, let off a reaction (annihilation radiation), creating images through the analogue device. David Kuhls legacy is seen through his initiation of tomographic imaging, and his inventions of the first emission tomographic instruments seen in 'Mark II', 'Mark III', and 'Mark IV'. His legacy will be continued through the late 20th century by other scientists and doctors who will further Kuhl's original ideas and theories to form the modern day PET imaging machines.

The Brownell Group and the First Demonstration of Annihilation Radiation

Annihilation radiation is an event in physical nuclear science where two particles collide. Typically, an electron meets with a type of positron, which is the antiparticle to an electron, creating an annihilation which projects photons (Charlton & Humberston, 2001). When considering how annihilation radiation contributes to positron emission tomography, radionuclides are important as they become radiotracers. When the radiotracers are injected into the body, they react with the antiparticles that exist in the body, creating imaging on the tomographic scale. This method was developed by Gordon Brownell when he developed the transaxial tomographic imaging technology (Phelps, Hoffman, Mullani & Ter-Pogossian, 1975).

Born in Duncan Oklahoma in 1922, Gordon L. Brownell (1922-2008) was a nuclear scientist and engineering professor at the Massachusetts Institute of Technology (MIT). After receiving his bachelor of science from Bucknell University and a PhD from MIT, Brownell was fascinated with nuclear science and it's application in medicine (MIT, 2008). During his career, Brownell played one of the major roles in developing positron imaging and emission tomography, as he established the Physics Research Laboratory at the Massachusetts General Hospital, dedicated to his research into the positron emission scanning (MIT, 2008). Most of Brownells career consisted of the exploration of positron technology in medicine, with contributions in multiple pieces of work supporting the use of a completely cylindrical scanners, specifically the scintillation parts in the scanner (Brownell, Burnham & Chesler, 1985).

Brownell reports on the first multiple detector positron imaging device in his 1999 history of positron imaging. He describes how in 1962 an increased need for resolution and commercial usability was necessary for advancement, and the best way forward was through a hybrid imaging option (Brownell, 1999). The PC-I was one of the first devices used as the tomographic imaging instrument with 2-dimensional rays for brain scans, and was further developed into a rotational PC-I which included "interpolative motion" to capture clearer and more resolute images (Brownell, 1999). The device could be understood in the PC-I and PC-II as a machine that has two outside panels that are positioned around the patient's skull, and a computer analog system that produced the tomographic imaging based on the reactions of the particles with their antiparticles (Brownell, 1999). Brownell's work was tested and featured by many organizations including the US Atomic Energy Commission Record of Inventions, who took the ideas and concepts that had been tested by Brownell and his associates and filed the first positron emission tomographic scanning device as an invention (Brownell, 1999).

One particular document that Brownell's work was featured in was the Medical Radioisotope Scintigraphy document out of the International Atomic Energy Agency. The document out of Vienna included a section where Brownell and his colleagues describe the developments in positron emitters through a cyclical production in terms of medical imaging (IAEA, 1968). The paper supported the idea of a type of nuclear energy that emitted low gamma waves for medical imaging and that the use of positron emission isotopes allows for an upper hand when comparing medical diagnostic imaging that was already in use during the 20th century (IAEA, 1968). The paper explains the use of annihilation radiation and how it is used to detect occurrences with particles. Through the device, annihilation can be detected on the field when the particles meet and pulse outward towards the detectors which in turn create reactions through the machine, creating an imaging system that shows where the annihilation happens (IAEA, 1968).

Michel Ter-Pogossian and The Modern PET Scan Machine Prototype

The development of the modern day PET scanner has evolved from different versions and has developed since the beginning of the 20th century when pharmaceutical agents were first discovered to be able to carry radiotracers to see and display organs within the body (Rubin & Rubin, 2017). Multiple scientists worked tirelessly researching and developing the nuclear medicinal technology to further enhance tomographic imaging. Alongside associates and colleagues of his, Michel Ter-Pogossian was able to develop the modern version of what we consider the PET scan. Ter-Pogossian's inventions and innovations propelled his career forward, cementing his legacy as the modern "father of PET" due to his extensive research on positron emission tomography and the use of radiotracers in patients (Science Museum, n.d).

Michel Ter-Pogossian (1925-1996), originally from Berlin in 1925, was an Armenian who fled from the Ottoman Empire during the First World War to Germany, and then later France where Ter-Pogossian would grow up (Wacker, 2018). Similar to Kuhl and Brownell, Ter-Pogossian was interested in science and experimentation from a young age, which would eventually lead him to a career in the nuclear medicine and positron imaging fields. After finishing his degree in France, Ter-Pogossian travelled to the United States in 1946 to pursue a PhD in nuclear physics from Washington University, Missouri (Wacker, 2018). During his career, Ter-Pogossian was invested in the development and usage of radioactive tracers (such as oxygen and nitrogen) in terms of imaging in a patient's metabolic system. Alongside a team of colleagues and other scientists, Ter-Pogossian was able to develop the first working prototype of a positron emission tomographic scan (Wacker, 2018).

Ter-Pogossian conducted multiple experiments and research tasks throughout his tenure, including research surrounding oxygen in autoradiographic techniques during the 1950's and an article in 1966 regarding nucleonics and the isotopes that are developed through a cyclotron (Rich, 1997). The hybrid model designed by Brownell was used as a base for Ter-Pogossian's future developments, as he was able to use oxy hemoglobins (that emerged in the 1970's) in an effective way by allowing for high doses to patients for optimal viewing and reporting, all while limiting patient exposure to radiation as the oxy hemoglobins being used had short half-lives (Rich, 1997).

Later on in 1975, Ter-Pogossian alongside his colleagues continued their work with the PET technology, and developed the PET III, which was a step up from the previous model as it included a full body scan where the detectors on the machine where in a hexagonal disbursement with up to eight (8) detectors per side (Rich, 1997). The schematic make up of PET III was introduced in the document that proposed the idea of full body imaging with the PET machine. The machine used a hexagonal shape, prompting the eight (8) detectors on each side to wait for the annihilation reaction of the particles within the patient. The patient is able to lay comfortably on a table with the machine at the head of the bed which can be adjusted for specific cases (see future chapters for more information on specific procedures regarding the modern PET scan)(Phelps, Hoffman, Mullani, Higgens & Ter-Pogossian, 1976).

These scans that transferred the radionuclides within the patient would later become the standard of PET imaging, and with the continual advancement of computers the PET machine was able to solve more issues that arose with tomographic imaging (Rich, 1997).

Abass Alavi's First PET Demonstration with Radionuclides

An honourable mention to Abass Alavi, who is noted as one of the first scientists to administer a radioisotope tracer in a human patient. After receiving his medical degree from the Tehran

University of Medical Sciences in 1964, Alavi moved to the United States and began working out of Philadelphia. Later in his career, Alavi began working the University of Pennsylvania where he would continue to study and develop radiology and nuclear medicine (Hoilund-Carlsen, 2018). Alavi was able to study under David Kuhl, and was the person to introduce a different type of radioisotope to the medical field called ((18)F-FDG) which is fluorine-18-fluorodeoxyglucose (Alavi, Moghbel, Alavi JB, 2014). This isotope made Alavi one of the most notable nuclear scientists during the late 20th century as it provided the PET technology with the ability for full body scans to be carried out in an effective and efficient manner that poses minimal harm to the patient (Hoilund-Carlsen, 2018). This led Alavi to a career in detecting disease and sickness within patients, and further providing a mainstream use and acceptance of nuclear technology in medicinal practices.

Conclusion

The positron emission tomography scanning machine has become one of the most mainstream devices used in medicine today in detecting disease and sickness. The scientists who have worked on and developed the PET scan tha we use everyday in modern times are responsible for the promotion of nuclear science in medicine. Kuhl's development of positron tomography, Brownell's advancement of annihilation radiation, Ter-Pogossian's model and prototype of the PET III, and Alavi's advancements of radiotracers and isotopes for patients have all marked nuclear medical history. These four scientists and doctors have created a machine that will continue to improve the medical field and physicians' equipment reliability, one scan at a time.

What are PET Scans?
By Jessica Henschel

Introduction

A positron emission tomography (PET) scan is a type of medical imaging that helps reveal how organs and tissues are functioning (Mayo Clinic, 2021). PET scans use a form of radioactive drug (tacer) to create 3D colour images to show tissue activity. This tracer may be injected, inhaled, or swallowed, depending on the organ that is being studied. Subsequently, the tracer drug collects in the areas of the body that have higher levels of chemical activity, which usually corresponds to areas of disease (Mayo Clinic, 2021). The radioactive material is made up of a radioactive isotope - an unstable atom that has excess radioactive energy - and is attracted to bodily material, usually glucose. This radioactive material releases positively charged particles (positrons) and a camera records the positrons; this is how the technician detects abnormalities (Canadian Cancer Society, 2021). The benefit of PET scanning is that it has been shown to detect disease before it is picked up by other tests, such as CT scans and MRIs. PET scans are done to identify a variety of illnesses, including cancer, brain disorders, and heart problems (Mayo Clinic, 2021).

Pictures from PET scans show information that assist doctors in diagnosing and assessing conditions. Different colours or degrees of brightness on a PET scan picture represent various levels of organ and tissue function (or lack thereof) (Virtual Medical Centre [VMC], 2017). PET scan images can be viewed in black and white (grey scale) or in colour (called "false color" viewing). Darker areas indicate lesser levels of cell activity and brighter areas indicate higher levels of activity. For example, cancerous cells tend to use more glucose than normal tissue and are growing much faster and therefore, show up as brighter spots on a PET scan because they use more of the radioactive tracer (VMC, 2017). Doctors will usually compare your PET scan results to other tests you've undergone to better understand your condition.

PET Scans for Cancer Detection

According to the Canadian Cancer Society (2021), PET scans are commonly used to identify many different forms of cancer. Cancer cells show up as bright spots on PET scans, as they have a higher metabolic rate than normal cells. Additionally, PET scans can be beneficial for: detecting cancer, revealing whether the cancer has spread, checking if a cancer treatment is working, and finding a cancer recurrence. There are many types of solid cancer tumours that can be detected by PET scans: brain, cervical, lung, lymphoma, prostate, thyroid, melanoma, and esophageal (Mayo Clinic, 2021). Due to PET scans' ability to detect early changes in cells, they are more useful for spotting cancer than other methods. CT scans and MRIs use special x-ray equipment or magnetic fields and radio frequencies to develop an internal picture of organs, tissues, and bone, but they do not necessarily show early changes or abnormalities in cells (Weaver, 2019). However, not all cancerous cells show up in PET scans and are often used in conjunction with other diagnostic methods.

Interestingly, PET scans have been shown to have particular efficacy in the diagnosis and treatment of non-small cell lung cancer (NSCLC) and small cell lung cancer (SCLC) (Ruysscher & Kirsch, 2010). This is especially important research, as PET scans have shown to be superior in early detection of lung cancer tumours for both NSCLC and SCLC (Ruysscher & Kirsch, 2010). The tracer thymidine kinase 1 with F-fluoro-l-thymidine (18F-FLT) has been studied extensively for lung cancer diagnostics and may provide information that assists with early intervention. PET scans also allow for more thorough staging and imaging of cancerous cells, thus avoiding unnecessary treatments. Performing a PET scan reduces the radiation treatment volume because of the avoidance of mediastinal lymph nodes - lymph nodes located in the thoracic cavity - and reduces the overall toxicity of cancer diagnostic testing. Overall, using diagnostic PET scans for common cancers can save time and resources, as it delays between diagnosis and treatment (Ruysscher & Kirsch, 2010).

PET Scans for Brain Disorders

Another common use of PET scans is for brain disorders, such as tumours, Alzheimer's disease, and epilepsy (Mayo Clinic, 2021). Alzheimer's disease is diagnosed using PET scans by measuring the uptake of sugar in certain areas of the brain. Brain cells that are affected by Alzheimer's usually use glucose at a decreased rate than healthy brain cells (Brazier, 2017). These scans also detect amyloid protein plaque in the brain, which are hallmarks of Alzheimer's. Since Alzheimer's leads to progessive loss of brain function and other cognitive abilities, it is important to catch it early and PET scans are the first step and are highly effective. One study found that PET imaging using florbetapir F 18 as a tracer was able to distinguish between 68 individuals with suspected Alzheimer's, 60 people with mild cognitive impairment, and 82 healthy elderly people with no signs of cognitive impairment (Mann, 2011). Another study examined PET scans using a different tracer - fluorine 18-labeled flutemetamol - among seven individuals with normal pressure hydrocephalus, which is a progressive brain condition that can mimic Alzheimer's and dementia. The PET scan was positively correlated with the biopsies taken to confirm Alzheimer's. Epilepsy is also commonly diagnosed using PET scans, as they can show the areas of the brain that are being affected by the seizures (Brazier, 2017). For both conditions, PET scans can help identify disorders and choose the most suitable treatment.

Pet Scans for Cardiac Conditions

Cardiac PET scans are also done to detect heart conditions, specifically heart disease (Mayo Clinic, 2021). They are used to diagnose coronary artery disease, detect areas of low blood flow in the heart, and examine any damage from heart attacks (American Heart Association, 2021). PET scans can also detect dead tissue and injured tissue that is still living. If the tissue is still viable, then patients may be able to undergo coronary artery bypass surgery or other forms of treatment. Cardiac PET scans also normally include the use of an electrocardiogram (ECG or EKG) - when small metal disks are placed on the patient's chest, arms, and legs to monitor heartbeat. The doctors will then take a baseline picture of the heart before the tracer is injected. Once the tracer is in the body, the PET scan and ECG/EKG will commence. A commonly used tracer for cardiac PET scans is either rubidium-82 or ammonia-13 and it is injected into the bloodstream and taken up by the heart (University of Ottawa Heart Institute, 2021). There is another method for performing cardiac PET scans and it can be separated into two distinct parts. First, a scan is done to measure the resting blood flow in the heart. Then the radioactive tracer will be injected into the bloodstream and the PET camera will detect the radiation released by the tracer in order to produce a picture (University of Ottawa Heart Institute, 2021). Secondly, a different radioactive tracer that contains glucose and fluorodeoxyglucose (FDG) is injected. This tracer is also taken up by the heart and another scan is done to show which parts of the heart may be damaged. The resulting image shows how the different parts of the heart use glucose. Since damaged or dead heart cells use little or no glucose, doctors can determine which areas of the heart may be affected from illness (University of Ottawa Heart Institute, 2021).

How to Prepare for a PET Scan

There are a number of things that an individual can do to prepare for a PET scan. By doing some of the items in the following paragraphs, you can reduce any concerns that may occur with a PET scan procedure (Canadian Cancer Society, 2021). Firstly, it is important to inform your doctor or PET scan technician:

- If you have ever had a bad allergic reaction to medications or food
- Of any history of other medical conditions, such as diabetes
- If you have been sick recently
- If you're pregnant or breast-feeding, or if you think you might be pregnant
- If you are claustrophobic or are afraid of enclosed spaces

Before the scan, you will be told not to eat or drink anything for 4-6 hours before you arrive for the procedure. It is also important that you avoid tobacco, caffeine, alcohol, or vigorous exercise for up to 24 hours before the scan. These factors may change the way the radioactive material being injected into your body reacts to the natural functions of your cells, as they may be affected by alcohol or caffeine, thus altering the results of your test (Canadian Cancer Society, 2021). You will also be told not to wear clothes with metal zippers, belts, or buttons on the day of the screening- you may also be asked to change into a hospital gown. If you have any kind of body jewelry, it is suggested that you remove it before you arrive for your appointment, as body jewelry may require assistance in removing and can delay the scan. Any other information will be provided to you prior to your test by the doctor and the clinic you are going to for your scan.

What to Expect During a PET Scan

Normally a PET scan is done on an outpatient basis at the nuclear medicine department of your nearest hospital or a specialized PET scan department (Canadian Cancer Society, 2021). When you arrive for your PET scan, the following events will occur. The PET scanner itself is a large machine that resembles a giant doughnut with a table attached to the opening- this is where you will be laying (Mayo Clinic, 2021). Sometimes, you will be put into a combined CT-PET scanner. Normally the entire visit can last up to three hours, depending on the area that is being examined and how smoothly the procedure goes. Once you have changed into a hospital gown and emptied your bladder, you will be given the radioactive drug tracer that is used to measure cell activity and is picked up by the PET scan. The tracer is either injected into the bloodstream, inhaled, or swallowed, depending on the area being examined (Mayo Clinic, 2021). If the drug is injected, you may briefly feel a cold sensation moving up or down your arm. Once the tracer has been put into your body, you will have to wait 30-60 minutes for the radioactive drug to be absorbed.

Once the drug has been fully taken up by the body, the procedure will begin. Normally, the scan itself will take anywhere from 45 minutes to 3 hours, depending on the area of the body being scanned and any issues that may occur (Canadian Cancer Society, 2021). The technician will instruct you to lay on a narrow, padded table that slides into the PET scanner. Once the exam begins, you will be slid into the PET scan and instructed to stay very still throughout the duration of the procedure. The table will then move through the doughnut shaped scanner. Detectors inside the PET scanner will pick up the signal from the radioactive tracer that is inside your body and a computer will then analyze the patterns. While this analysis is being done, the computer is creating a 3D colour or greyscale image of the area of your body being scanned- this is why it is imperative to stay as still as possible (Canadian Cancer Society, 2021). Some individuals start to feel an overwhelming sense of anxiety while inside the scanner. Make sure to alert your medical staff if this occurs. If you are prone to claustrophobia or anxiety of small spaces, it is best to alert your technicians before the procedure, as they can give you medication to help you relax and make the process go smoothly (Mayo Clinic, 2021).

One the scan is done, you will be slid out of the PET scanner and escorted back to the area where you can change into your everyday clothes. The radioactive tracer will eventually pass through your system via stool or urine (Mayo Clinic, 2021). In order for the tracer to fully exit your body, it might take a few hours or a few days. It is important to stay hydrated and drink lots of fluids to assist your body with secretion of the drug. You should be able to carry on your day as usual, as PET scans do not impair any kind of body functioning. However, it is always a good idea to consult your medical professional after the scan to ensure you are taking the right steps.

Risks and Side Effects

Since the dose of radioactive tracer is different for each nuclear medical imaging procedure, the potential risks and side effects can vary (Canadian Cancer Society, 2021). Bruising and swelling at the site of radioactive tracer injection is normal. There is a risk that the tracer will leak outside the vein and it can cause more swelling and pain than normally is expected. The dosage of radioactive material depends on the area of the body that is being scanned. Normally, the dose of radioactive material that is being consumed or injected into the body during a PET scan is small and you are exposed to very low levels of radiation during the scan itself. The tracer is essential glucose with a radioactive component attached to it. This makes it very easy for your body to dispose of the radioactive material, even if you do have a history of kidney disease or diabetes (Krans, 2018). Therefore, the risks are generally minimal and the benefits of having a PET scan outweigh the risk of exposure to small amounts of radiation.

Despite the low risk associated with PET scans, there are some possible side effects for individuals with certain conditions or allergies (Krans, 2018). It is possible to have an allergic reaction to the tracer, especially with those who are allergic to iodine, aspartame, or saccharin. Those who are unable to have an iodine tracer will alternatively be given one that contains diluted barium sweetened with saccharin. Those who are most likely to have a negative reaction include those with:

- A history of allergic reactions to PET scans
- Asthma
- Allergies
- Heart disease
- Dehydration
- Blood cell disorders including sickle cell anemia, polycythemia vera, and multiple myeloma
- Kidney disease
- A drug regimen that includes beta-blockers, nonsteroidal anti-inflammatory drugs (NSAIDs), or interleukin-2 (IL-2)

Additionally, radiation is not considered safe for those who are pregnant or breast-feeding. Radiation can severely impact a developing fetus and should be avoided if you believe you might be pregnant (Krans, 2018). If you are breastfeeding, the department you are getting scanned at will let you know how many days you should go before breastfeeding again to allow for the radioactive material to exit your system and to avoid any harm that may come to your child.

What Science is Involved in PET Scans?

By Hassan Khan

Introduction

Over the past year, the COVID-19 pandemic has changed the way individuals function, with the closure of businesses, schools, and offices. However, during this time, vaccines were developed at a faster rate than ever before in a rush to vaccinate most of the population and reach a certain level of immunity. Without the cutting edge technology and the minds of researchers who used their knowledge of the sciences (chemistry, biology, and physics), this would not have been possible. The interdisciplinary nature of these sciences helps to explain how PET scans function. Chemistry is the study of how structure and properties of molecules help to explain reactions (Harvard University, 2021). Biology is the study of living things, specifically how proteins, enzymes, and genes are involved in chemical and biological processes (Harvard University, 2021). Physics dives into the fundamental laws of nature by looking at atoms, molecules, nuclei, and nuclear particles (Harvard University, 2021). This chapter will go into detail of these three branches of science that are important in their use of medical applications (i.e. imaging), most notably PET scans.

The Importance of Fluorine

Fluorine is widely known in chemistry to be the most electronegative element in the periodic table (Alauddin, 2012). In fact, most elements can be strongly covalently or ionically bonded to it (Alauddin, 2012). In organic chemistry, it forms one of the strongest bonds when bonded with carbon, and for this reason, it is widely used in pharmaceuticals (Alauddin, 2012). Despite being polarized, the carbon-fluorine bond is still stable due to electrostatic interactions between the two atoms (Alauddin, 2012). About 1/5th of pharmaceutical companies use Fluorine in their drugs, such as 5-fluorouracil, flunitrazepam, fluoxetine, paroxetine, ciprofloxacin, and fluconazole (Alauddin, 2012).

In addition, fluorine has many isotopes such as fluorine-19, fluorine-18, fluorine-17, fluorine-20, and fluorine-21 (Alauddin, 2012). All of these isotopes except fluorine-19 are radioactive and have short half-lives (Alauddin, 2012). The fluorine-18 is particularly radioactive due to its shortened half-life and is therefore, more commonly used in the scientific community (Alauddin, 2012). Using a cyclotron (produces radioactive isotopes), fluorine-18 is made by proton irradiation of oxygen-18 (a stable isotopic form of oxygen) (Jacobson et al., 2015). A cyclotron is particularly useful for creating short half lives of isotopes by accelerating protons (Anosh, 2016). This device works on the principle that magnetic forces are able to bend moving charges (protons) (Anosh, 2016). In a cyclotron, there is a gap between two metal components that reverses the electric field (Anosh, 2016). If a proton is placed near the centre, it accelerates away from the centre's electric field (Anosh, 2016). The field then reverses, so the proton accelerates back towards the centre, gaining kinetic energy (Anosh, 2016). Once the proton has enough energy, it is able to reach the nucleus of the oxygen-18 isotope and cause proton irradiation (Anosh, 2016). When fluorine-18 is interacting with the liquid form of oxygen-18, an aqueous solution of fluorine-18 (fluoride ion) is formed (Jacobson et al., 2015). When interacting with the gas form of oxygen-18, a gas form of fluorine-18 is produced (Jacobson et al., 2015). The fluorine ion is used as a nucleophile (donates electron pair), while the gas is used as an electrophile (accepts electrons) (Jacobson et al., 2015).

The difference between these two forms is their radioactivity with the ion having greater radioactivity and the gas having less (Jacobson et al., 2015). This radioactivity becomes important in ligand-receptor interactions (of molecules) as the fluorine-18 gas has lower activity and needs a carrier such as fluorine-19 to remove it from the target (Jacobson et al., 2015). As such, the final mass of the radioactive tracer is greater than the fluorine-18 ion (Jacobson et al., 2015). This causes increased receptor saturation and reduced PET signal from binding (Jacobson et al., 2015). Therefore, fluorine-18 is most widely used in nuclear medicine because of its increased yield from the cyclotron and high radioactivity levels (Jacobson et al., 2015).

The Discovery of the FDG molecule

The scientific discovery process of the FDG (fluorodeoxyglucose) molecule began when George de Hevesy, a chemist and physicist used radioactive tracers to unveil the secret molecular processes happening within the body (Naider et al., 2011). The main challenge was to take naturally occurring molecules and convert them into radioactive molecules while also being easily accepted into a human's bloodstream (Naider et al., 2011). It did not take long for many researchers to try and figure out what would work, and eventually the FDG molecule was discovered which was later known to be the "molecule of the 20th century" for PET scans (Naider et al., 2011). Later in the chapter, FDG will be discussed in greater detail with respect to the biological processes it plays an important role in.

When a PET scan is being performed, the tracer given to the patient is a drug also known as a radiopharmaceutical that contains a radioactive positron emitting atom (Alexander et al., 2018). This substance allows us to measure the different functions of the body including blood flow, oxygen level, and sugar metabolism (Alexander et al., 2018). Radiopharmaceuticals are composed of two things, the radionuclide and the drug itself (Alexander et al., 2018). The radionuclide is an isotope that contains a short half-life but remains traceable in the body ("Radiation in Biology", 2020). It contains a fraction that emits radiation when detected by the PET scan (Alexander et al., 2018). Some examples of common radionuclides include Carbon-11, Copper-64, Fluorine-18, Krypton-79, Nitrogen-13, and Oxygen-15 ("Radiation in Biology", 2020). The drug is a chemical fraction that assists in the process of leading the radiopharmaceutical to the desired location within the body (Alexander et al., 2018).

The use of FDG in PET Scans

The most commonly used radiopharmaceutical used in PET scans is FDG (Alexander et al., 2018). FDG is a structurally similar molecule to that of glucose that is used as a marker for metabolic activity (Alexander et al., 2018). FDG is generally found in those areas of the body where glucose consumption is high including the kidneys, brain, and cancer cells (Alexander et al., 2018). The process of FDG begins by entering the cells and going through the various membrane receptors where it follows the same metabolic pathway of glucose (Alexander et al., 2018). FDG is then phosphorylated by hexokinase by glucose-6-phosphatase and is converted into 18F-FDG-6 phosphate (Alexander et al., 2018). In the presence of normal cells, this process will not continue any further so the concentration remains balanced, however in tumor cells, the concentration remains much higher (Alexander et al., 2018). With this in mind, the difference in these concentrations allow us to properly assess and diagnose patients with the use of PET scans (Alexander et al., 2018). It is important to note that the radioactivity of FDG subsides relatively quickly, so no harm is done to the patient receiving it (Naider et al., 2011).

An FDG molecule is made by combining a positively charged proton to an oxygen molecule (Naider et al., 2011). As it loses its radioactivity, one of the protons is converted into a neutron so it becomes more stable (Naider et al., 2011). However, another positive particle known as a positron is also released that collides with a negatively charged electron (Naider et al., 2011). This collision process results in energy being emitted leading to the formation of gamma rays which are detected by scanners (Naider et al., 2011). To summarize, gamma rays are packets of energy emitted by the protons of the radionuclides nuclei as a result of radioactive decay (Arpansa, n.d.).

Specifically, PET scan images are generated by scintillation detectors which contain detection elements within them to facilitate the process (Schmitz et al., n.d.). The scintillation elements are able to give out visible or ultraviolet light after they have collided with the photons from the FDG molecule (Schmitz et al., n.d.). The photo detectors then recognize and measure the scintillation photons leading to the generation of a PET scan image (Schmitz et al., n.d.). When PET scan images are being generated by the detector, they tend to be more blurry and have less detail when compared to CT (Computed Tomography) and MR (Magnetic Resonance) images, due to the lower number of photons being emitted (Schmitz et al., n.d.).

The Physics behind PET Scans

Furthermore, there are four main properties of Scintillators that are important for the best possible PET scan image (Schmitz et al., n.d.). They include stopping power, decay constant, energy resolution, and light output (Schmitz et al., n.d.). A light beam's stopping power is inversely proportional to the distance it travels before depositing its energy (Schmitz et al., n.d.). Shorter distances are better because less energy is lost and there are more interactions with the photons prior to reaching the detector (Schmitz et al., n.d.). Decay constants measure how long the scintillating elements can remain active, and shorter decay constants are better for counting more photons (Schmitz et al., n.d.). With energy resolution, energy variance over energy is taken into account, so that the measurement has fewer fluctuations (Schmitz et al., n.d.). In a PET scan, the light output is the number of scintillation photons that are emitted from each photon; more photons result in better resolution of the scan image (Schmitz et al., n.d.). The resolution of many PET scans are (6-8)3 mm3, but some clinical brain scanners are known to approach much higher resolutions (Gambhir, 2002). An image can usually only be captured if hundreds of millions of cells are gathered close to the tracer (Gambhir, 2002). It is uncertain how many cells can be imaged due to many factors, including the amount of tracer injected into the surrounding tissues (Gambhir, 2002). All isotopes used produce gamma rays of the same energy, so two molecular probes of different isotopes installed together will not be detected by the PET scan (Gambhir, 2002). That is why studies of multiple molecular events usually employ separate molecular probes to allow for isotopic decay (Gambhir, 2002). To summarize, PET scan images will show gamma ray events of the same energy with the colour aspect referring to the concentration of the isotope (Gambhir, 2002).

In addition, PET scan images are usually reconstructed because the images first obtained are not exactly a tomographic image (Cherry & Gambhir., 2001). These reconstructed images are done through CT (computed tomography) in order to get detailed images (Cherry & Gambhir., 2001). Based on measured gamma ray events, the signal-to-noise level of the reconstructed images will depend on the number of gamma rays composing the data set (Cherry & Gambhir., 2001). PET scans are most commonly performed with PMTs (photomultiplier tubes) containing cathodes that absorb photons and use them to create electrons that are amplified (Schmitz, n.d.). The energy current generated is proportional to the number of scintillation photons, as well as to the amount of energy deposited in the scintillation elements by the PET photons (Schmitz, n.d.).

During a dynamic PET study, volumetric images are taken, showing the radionuclide moving over time (Cherry & Gambhir., 2001). In the case of small animals, these images can showcase the entire body (Cherry & Gambhir., 2001). As well, a lot of useful biological information can be derived from PET data by using tracer kinetic models (Cherry & Gambhir., 2001). With tracer kinetic models, the amount of radioactive molecules can be measured and used to determine metabolic rates, accumulation rates, and the binding characteristics of the molecule attached to the target of interest (Cherry & Gambhir., 2001).

Properties of PET Tracers in the Central Nervous System

There are four main properties that a PET tracer must carry when embedded into an individual's central nervous system (Hietala, 2009). These are: specific and selective receptor binding profile, appropriate binding affinity and receptor association, optimal lipophilicity (ability to dissolve in fats and oils), and low metabolism (Hietala, 2009). The tracers should be specific and selective to the drug receptors, so that the PET scan is able to capture the area of interest (Hietala, 2009). If the binding affinity is too low, this will lead to problems in the PET scan image (Hietala, 2009). If it is too high, then the binding may become delivery-dependent in high density areas (Hietala, 2009). However, for capturing a good PET scan image, high affinity is necessary for detecting low densities of receptors (Hietala, 2009). Solubility of lipids should be at a standard level, as too low solubility can impair blood-brain barrier penetration (prevents drug uptake to the brain) (Hietala, 2009). Lastly, during a PET scan study, metabolism rates should be low and the formed metabolites should be hydrophilic in order not to interfere with the signal from the brain and bias modelling (Hietala, 2009).

Novel Technique in Labelling Compounds

During 2017, a discovery was made when scientists at the Department of Energy's Lawrence Berkeley National Laboratory uncovered a new technique for attaching chemical tracers (Roberts, 2017). By using the mechanism, CF3 (trifluoromethyl) compounds can be made and attached to other compounds (Roberts, 2017). This could open up new possibilities regarding chemical reactions occurring in the presence of catalysts, including gold and copper (Roberts, 2017). This could also lead to the formation of new radioactive tracers that can be used with

PET scans (Roberts, 2017). There are drug companies who are increasingly more interested in this invention of CF3 compounds (Roberts, 2017). This can help make medications more effective and last longer (Roberts, 2017). Some examples of drugs containing CF3 compounds are Prozac (antidepressant HIV drug) and Celebrex (anti inflammatory) (Roberts, 2017).

As always with tracers, the CF3 compounds should be attached with the radioactive fluorine-18, so that it is detected by PET scanners (Roberts, 2017). A lead author working on this study, Mark D. Levin, mentioned, "This research allows us to make new classes of radioactive tracers that had been too difficult, or even impossible, to prepare previously" (Roberts, 2017). It has been known for a while that the CF3 groups have medical applications, but they were very difficult to use previously (Roberts, 2017).

Conclusion

The COVID-19 pandemic changed the way society functions with social distancing, face masks, and the closure of many businesses. However, the ability of researchers to use their minds and scientific knowledge played a huge role in the production of vaccines. It would have taken much longer to reach that stage, if science was not a key factor in each country's decisions. Likewise, PET scans are also playing an important role in the medical community by helping identify tumors/cancer for patients affected by these conditions. Through this chapter, it is evident that the science behind it is just as crucial to understand as well. The scientific discovery and chemical/biological processes of the FDG molecule and its role in PET scans has played a huge part in important applications. As with the physics behind PET scans, photons, gamma rays, and energy are integral for constructing the final image from a PET scan. Similarly, there are key properties to PET tracers in order to produce a successful scan image. There are also new discoveries being made with PET imaging such as in the case of CF3 compounds adding a new layer to gold chemistry. Due to these remarkable concepts and innovations, scientists and researchers are able to offer patients such medical applications.

What Have We Learned About PET Scans in Recent Medical History?

By Ruchira Nandasiri

PET/CT Combined Imaging

The monitoring of the diseases through imaging has its own positive outcomes including the staging, monitoring and detection of the diseases (Purz et al., 2014). Thus, methods including positron emission tomography (PET) allows the visualization of body functions with the detection of radiopharmaceuticals at lower levels which benefits both the medical and pharmaceutical industries (Purz et al., 2014). PET/computed tomography (CT) system began its functionality in 1998 and kept its pace throughout the years (Townsend & Beyer, 2002). The success behind the combined PET/CT lies with fluorine 18 (18F) fluoro-deoxyglucose (FDG) being highly synergistic in tumor therapy and monitoring applications (Von Schulthess et al., 2006). The combined FDG-PET/CT brings into the table a high sensitivity of PET, which enables the detection limits in very low concentration (picomolar) and the high specificity of CT which provides relatively higher accuracy than either technology alone (Eyuboglu et al., 2021). Indeed, the combined PET/CT provides precise locations of the damages with increased specificity (Hany et al., 2002). As a result of the advancements in the combined imaging technology the annual examination rate for FDG-PET/CT increased at a rate of 7% in the year 2018 for the United States (Eyuboglu et al., 2021). In traditional PET scanners data were obtained via radioactive transmission source swiveling around the patient similar to a CT scan with a lower photon flux. However, a combined PET/CT facilitates 25% more faster data acquisition than the CT alone

(Hany et al., 2002). Furthermore, current data indicates that combined PET/CT is highly specific and sensitive and has higher accuracy in detecting the tumors than the separate PET and CT systems alone (Von Schulthess et al., 2006). It has been reported that conventional PET imaging is not suitable for tumor staging due to its lack of identity on tumor confines anatomically, whereas the PET/CT showed promising results on tumor staging of lung cancer patients (Lardinois et al., 2003). Recent developments in the industry warrants highly efficient PET components with different detector materials which would result in faster image acquisition with higher spatial resolution (Von Schulthess et al., 2006).

PET/MRI Combined Imaging

A recent development in the combined functionality effect of PET/magnetic resonance imaging (MRI) has lower radiation onus than the PET/CT. Moreover, combining both PET with MRI advances the imaging quality with high resolution and contrast, thereby improving the image fusion and anatomic resolution. Furthermore, the combined PET/MR would provide new perspectives for non-invasive imaging techniques for clinical management and research (Purz et al., 2014). As the combined method of PET/MR has lower invasiveness this could be recommended as a method for monitoring and detection of the pediatric diseases due to its lower radioactivity and the reduction of the exposure to repeated diagnostic imaging sessions (Purz et al., 2014). Additionally, the combined effect of PET/MRI provides corresponding anatomical, physiological, and functional details related to the brain. The first ever feasibility study on concurrent attainment of the human data was conducted in the year of 2007 using a 3-T MRI Trio system (Siemens Healthcare Inc.) (Schlemmer et al., 2008). This was further developed into a PET/MRI scanner full body imaging in the year of 2010 (Biograph mMR; Siemens) (Drzezga et al., 2012).

Application of PET in Medical Industry

This combined impact of PET/MRI allows the acquisition of both spatial and temporal correlation of the signal thereby creating novel applications in neuroimaging, neurologic and psychiatric research (Catana et al., 2012). Furthermore, the combined PET/MRI further facilitates the in-vivo valuation and cross-correlation of numerous neuropsychologic parameters including changes in hemodynamics, cerebral blood flow (CBF), cerebral blood volume (CBV), and oxygenation, ischemia, necrosis, and apoptosis (Catana et al., 2012). Moreover, many other disorders pertaining to the central nervous system over time might also be improved with the combined PET/MRI imaging technique (Catana et al., 2012). Moreover, with the application of this combined effect the frequency of sedation or anesthesia could be reduced significantly (Purz et al., 2014). Recently, the MRI was substituted by functional magnetic resonance (fMR) which combines both anatomic and bio-physiologic information with diffusion-weighted imaging (DWI) where it signifies the dynamic contrast–enhanced imaging representing the blood vessels associated with tumors (Purz et al., 2014). However, often the incorporation of fMR with PET may lead to longer scanning durations which may not be a feasible option with the pediatric patients. It was found that combined PET/MRI would enhance the performance while improving both qualitative and quantitative information. Further, the combined impact of PET/MRI imaging would improve the accuracy of the PET with jointly validating the MRI techniques in-vivo (Catana et al., 2012). One major advantage of this combined PET/MRI is that it reduces both the number of examinations and time for neuropsychiatric patients while increasing the patient comfort (Catana et al., 2012).

Application of PET in Pediatric Studies

In children, both non-Hodgkin and Hodgkin lymphoma account for over 10% of the tumors with relatively higher curing rates. Thus, the diagnosis is important, and imaging has a key role in determining the stage and the magnitude of the tumor (Buchbender et al., 2012; Kluge et al., 2013; Montravers et al., 2002). Furthermore, Montravers et al. (2002) indicated that 18F-fluorodeoxyglucose (FDG) PET has more sensitivity than the conventional imaging techniques where it illustrates more lesions. Further, this novel technique has more accuracy in monitoring the response to chemotherapy and prototyping the residual nodes which are common with Hodgkin lymphoma. Moreover, in another study conducted by London et al. (2011) demonstrated that the sensitivity and specificity of 18F-FDG PET/CT imaging (95.9% and 99.7%) was much higher than the conventional imaging (70.1% and 99.0%). Additionally, the evaluation of bone marrow intrusion is another important factor in determining the stage of lymphoma, whereas 18F-FDG PET reported to have higher sensitivity and specificity (Purz et al., 2011). In another study conducted by Wu et al. (2012), proved that combined 18F-FDG PET/CT imaging and MR had similar sensitivities (90%) in diagnosis of lymphoma while the specificity of MR was much lower than the combined imaging. Hence, the misinterpretations of post therapeutic changes including bone marrow edema, necrotic tissue, or age-dependent changes can be minimized using the combined imaging technique (Wu et al., 2012). While conventional imaging has many limitations in identification of viable tumors and fibrosis PET is often awarded due to its ability to distinguish between viable tumor and fibrosis in residual masses (Purz et al., 2014).

Application of PET in Neuroimaging

Apart from the identification of lymphoma tumors, PET has other important contributions towards the identification of neurologic neoplastic and nonneoplastic diseases. Furthermore, the role of PET in detecting biochemical and molecular changes prior to structural changes or clinical symptoms has many advantages in neuroimaging (Garibotto et al., 2013; Singhal, 2012). The application of PET tracers using 18F labelled amino acids (O-(2-18F-fluo-roethyl)-L-tyrosine) and 11C labelled methionine in brain imaging has been widely established (Catana et al., 2012). On the contrary, other tracers including proliferation marker 3′-deoxy-3′-18F-fluorothymidine or 15O-labeled water and oxygen as perfusion markers are applied with PET and used in clinical research (Catana et al., 2012). Furthermore, the combined PET/ MR treatment imaging provides accurate details of functional, molecular, and morphologic interactive pathways with enhanced in-vivo measurements (Purz et al., 2014). The combined PET/ MR imaging could be applied towards the pediatric neuroimaging since the ordinary type of solid malignancies in children are associated with the tumors of the central nervous system (Hipp et al., 2012). Acquisition of precise data on tumor biology and tumor response after anti-tumoral therapy with combined PET/MR aids in guiding surgeries, radiation therapies and biopsies (Catana et al., 2012).

Application of PET in Unknown Type Inflammations and Fever

Both infections and non-infectious inflammatory diseases are oftentimes associated with fever or unknown type of inflammations. However, the identification of these diseases is challenging due to the lack of morphological changes in its early stage. Thus, early detection and localization of these unknown types are important in further diagnosis (Balink et al., 2009; Jasper et al., 2010). Many studies have reported that 18F-FDG PET/ CT has the ability to diagnose these unknown types of infections and non-infectious inflammatory diseases (Balink et al., 2009; Jasper et al., 2010). Furthermore, a study conducted with children who were diagnosed with unexplained inflammatory signs at the latter stage (n = 37, 54%) with 18F-FDG PET and PET/ CT demonstrated that PET was diagnostically helpful (Jasper et al., 2010). Additionally, Berthold et al. (2013) investigated the 18F-FDG PET imaging technique as a tool for identification of inflammatory bowel disease in children and found the 18F-FDG PET imaging could be used as a method for determining the degree of inflammation where endoscopy is inaccessible. Hence, the combined approach of PET/MR may have potential benefits in determining and monitoring inflammatory bowel disease where corresponding imaging is required for the characterization of the injuries of the intestine. Additionally, many new PET tracers including cyclooxygenase 2 inhibitors are also under investigation for the treatment of neuroinflammation. Thus, a hybrid whole-body PET/MR is recommended for the diagnosis of infectious and inflammatory diseases compared to PET/CT or PET alone (Purz et al., 2014).

Application of PET in Molecular and Cellular Imaging

Molecular and cell imaging is another important field in both the biotechnological and medical industry which has higher potential with progressive growth. The application of PET/MRI in the cell imaging field would be a valuable source for innovative treatment strategies including therapeutic effects of targeted gene transfer, stem cell transplantation, and cell replacement methodologies. The level of gene expression and the impact of transferred genes on the tumor growth and metabolism can be functionally determined via the combined PET/MRI imaging (Jacobs et al., 2001). Cell replacement has been recommended in many cases as a method of treatment to cure many neurologic disorders (e.g., ischemic stroke). To proceed with the treatments, it is often necessary to observe and follow the movement of grafted stem cells (Hoehn et al., 2007). Moreover, this combined imaging technique can be applied to determine the cell viability and differentiation after cell transplantation. Monitoring the cell movement and its viability across the cells into functional networks via combined imaging technique of PET/MRI holds promising results over fetal graft transplantation for many neurologic diseases as well as new molecular neuroimaging applications with potential scientific benefits (Hoehn et al., 2007).

Evaluation of myocardial viability through PET

The assessment of myocardial viability is vital in managing the ischemic cardiomyopathy which leads to ischemic heart diseases which accounts for the leading cause of death worldwide (Khalaf et al., 2019). The viability of ischemic cardiomyopathy is conducted in many non-invasive methods including single-photon emission computerized tomography (SPECT), PET, dobutamine stress echo and cardiac magnetic resonance assess (Jamiel et al., 2017). However, recently PET was preferred over SPECT due to its greater image quality and high data accuracy. Furthermore, PET uses 18F-fluro-deoxyglucose (FDG) which creates elevated positive and negative predictive values, greater spatial and contrast solution, reduced radiation exposure, quicker protocol, and its capability to scan obese patients with less attentiveness (Khalaf et al., 2019). This was evident with the results from a study conducted Di Carli et al. (1998) where patients who underwent a PET guided revascularization demonstrated relatively higher recovery rates in heart failures including angina. Furthermore, these patients had better survival rates than the patients on the medical therapy (Di Carli et al., 1998). Additionally, PET/MRI combined imaging has huge potential in receiving the necessary information related to cardiac malignant tumors and occult tumor sites which helps in surgical planning and postoperative surveillance (Khalaf et al., 2019). Thus, the application of this hybrid imaging system in myocardial viability is still under investigation and more research is required with regard to its clinical outcomes, cost-effectiveness, technical achievements and optimization of the protocols.

Evaluation of Skeletal Scintigraphy Through PET with 18F-Fluoride

Years before the PET was invented 18F-labeled NaF was applied in the skeletal imaging as an excellent source of radiopharmaceutical. Further, this 18F-labeled NaF was used in skeletal scintigraphy and was approved by the FDA in 1972 for clinical use (Grant et al., 2008). However, with the advancement in the science in early 1990's this 18F-fluoride was used coupled with PET as a model for developing whole body PET imaging (Hoh et al., 1993). The primary application of 18F-fluoride imaging was on bone tumors whereas it was further used to detect skeletal metastasis in the patients with primary tumors (Grant et al., 2008). In a comparative study using 18F-fluoride PET discovered over 90 metastatic lesions in 15 patients whereas 99mTc-MDP planar scintigraphy was only able to detect over 40 metastases (Schirrmeister et al., 1999). Further, researchers have found out that compared to skeletal SPECT, 18F-fluoride PET can be applied to detect the alteration in the bone marrow created by the stress induced intense sports. Moreover, skeletal 18F-fluoride PET is applied towards determining the bone viability after trauma or reconstructive surgery signifying its use in skeletal scintigraphy (Grant et al., 2008). Apart from that many investigations were conducted to assess the bone turnover using 18F-fluoride PET. The researchers have found that quantitative assessment of 18F-fluoride PET as a non-invasive measure to determine the bone turnover correlating to its bone histomorphometry (Cheng et al., 2013). Additionally, the imaging time required for 18F-fluoride PET is significantly lower than the other imaging techniques related to skeletal scintigraphy. On average it takes around 15-30 minutes for 18F-fluoride PET imaging, whereas 99mTc-MDP SPECT would take around 1 hour, and 99mTc-MDP would take over 3-4 hours (Grant et al., 2008). The high specificity and rapid clearance from the blood pool with minimal protein binding ability holds promising results in 18F-fluoride PET imaging which may have potential impact in the medical industry in near future. Furthermore, with confirmed clinical data 18F-fluoride PET imaging would provide beneficial assessment of metabolic bone disorders including renal osteodystrophy, osteoporosis, or Paget's disease.

Conclusion

As per conclusion we could say that PET imaging has a huge potential towards the medical industry. Even though there are still misconceptions about the PET with the advancement in the science and the application of the new hybrid PET imaging systems would provide a better understanding of the patients and would further reduce the exposure time to radiation and other harmful rays significantly. Novel data further explicit that this PET imaging can be advanced into other industries including, food science, geoscience and engineering. Thus, the major drawback in the current radiology industry is the lack of technical experts and the huge capital investments associated. Hence, in recent times these new hybrid systems would further reduce the capital investments and the costs associated with the digital imaging.

What are the Various Uses of PET Scans in Human and Animal Sciences?

By David Supina

Introduction

PET scans are used for diagnosing a number of different conditions, like Alzheimer's or various heart diseases, but they are also used for monitoring the progress of a condition, perhaps while a patient is undergoing treatment (Brazier, 2017). Before we get into what PET scans are typically used for, it may be helpful to examine how PET scans contrast with other scanning methods that are used for similar ends.

PET Scans Versus MRI Versus CT Scans

In a PET scan, a person is typically injected with a radioactive tracer that will help identify the function of the brain, heart and other bodily functions (Krans, 2018). It is used more often to examine organs rather than bones (Weaver, 2019). Often problem areas are identified by their greater or lesser reaction to the radioactive tracer than other, normally functioning portions of the body (Brazier, 2017). It does use radiation, but not nearly as much as CT scans or X-rays do (AICA Orthopedics, 2019). These scans, however, can be less detailed on their own compared to combination PET-CT or PET-MRI scans (Weaver, 2019).

A CT (computed tomography) scan functions somewhat like an X-ray, but offers a fuller, three dimensional image. It takes a series of X-ray scans, each from a different angle, to gradually build a three dimensional image of the body. It can be used to look for cancer or problems with bone joints, internal bleeding or heart disease, tumours or blood clots (Independent Imaging, 2021). These scans can be quicker (five to ten minutes) and significantly less expensive than other tests. There is a moderate amount of radiation involved, which does involve some risk as a result, especially if multiple CT scans are wanted in a short period of time (Medical Imaging of Fredericksburg, 2019).

An MRI (magnetic resonance imaging) also takes detailed images, but instead of using X-rays, it uses magnetism, radio waves and computer technology in order to compile its images. Since it is using a different basis than PET scans or CT, there is no radioactivity used (Independent Imaging, 2021). These scans are often used if there is not enough detail after a CT scan has been done, and are often very good at diagnosing a variety of injuries, such as broken bones or torn ligaments. It can also detect blood clots or tumours (AICA Orthopedics, 2019).

Of course in practice, PET scans are often combined with MRI or CT to create even more robust images (Krans, 2018). While MRI or CT scans are often good at catching the structural changes in a body, PET scans are more adept at catching chemical and physiological differences. Stated another way, CT or MRI may give you good information about where abnormalities are located, but not about precisely how the organs may be functioning; the latter is more the domain of a PET scan. The complementary nature of the information gained is why they are often done together. These scans can often be done concurrently, and do not always require separate tests (Brazier, 2017).

A PET scan can give information at a cellular level for the body. This gives it uses that other kinds of scans do not, and allows them to potentially catch some diseases and bodily dysfunctions at a comparatively early stage. While there are a wide number of possible disorders that can be caught by a PET scan, there are several areas of assessment that PET scans offer for human beings. A PET scan may be used to assess blood flow, glucose metabolism or oxygen usage. Some key categories of use for PET scans include monitoring cancer, heart health and brain function (Krans, 2018). We will go through each in a little more detail.

Cancer

PET scans are often used to detect whether or not there is a presence of cancer, but that is not its only use. It can also be used to see if cancer has spread throughout the body, the efficacy of cancer treatment or to see if cancer has recurred. This is unfortunately not a process without error, however, as it is possible for cancerous tumours to not appear on a PET scan, or for benign parts of the body to appear to be cancerous (Krans, 2018).

The reason that PET scans work for identifying cancerous cells is that there is a difference in the metabolism of cancerous as opposed to benign cells; the rate of metabolism is higher in cancerous cells (Stanford Health Care, 2020). This means that on the actual image produced, cancerous cells will show up as bright spots. PET scans can also be helpful for medical doctors to determine whether or not a biopsy would be appropriate (to determine the type of cancer) or facilitate in the planning of radiation treatment (Cancer.Net, 2020). These scans may also catch a problematic tumour sooner than other methods of monitoring cancer (Brazier, 2017).

When looking for cancer, it is a combination PET/CT scan that is often used (Cancer.Net, 2020). There are several reasons for this. By doing them at the same time, there is greater accuracy and less error because you do not have to factor in the images being taken at different times from different angles (Radiologyinfo.org, 2019). It is also much more convenient to have the two scans done at once rather than having to worry about two different scans at two different times.

There is also the potential for information to be gained on how to treat cancer from PET scans being used in animals, which is something we will take a look at a little later. For now, it's worth simply observing that humans and animals are not fundamentally different things, and the lessons we learn from the diagnosis and treatment of non-human animals can often be applied to us.

Heart

Scans can be ordered of the heart due to a variety of possible worrisome symptoms, such as shortness of breath, tightness in the chest and arrhythmia, which is also known as an irregular heart beat (Krans, 2018). A PET scan is not a one-stop shop for diagnosing heart and cardiovascular problems, and it is expensive, so other tests may be ordered beforehand, such as an echocardiogram or a cardiac stress test (Krans 2018). If those tests do not provide enough information, a PET scan of the heart may be in order.

A PET scan can often reveal the level of blood flow to or from the heart. Tissue that is operating normally, IE healthy tissue, is more capable of taking on the radioactive agent used in PET scans than unhealthy tissue, or tissue that has less blood flow (Krans, 2018). There can be a variable degree of function indicated by PET scans, so the imaging does not fit the binary of does or does not have blood flow. Additionally, the PET scan can indicate if there have been parts of the heart that have been damaged, or may be diseased (Brazier, 2017).

There are a number of benefits possible to undergoing a PET scan for the heart, mostly revolving around giving doctors the chance to accurately diagnose the condition of the patient. As mentioned, it is possible for the doctor to see an accurate map of the blood flow in and around the heart, and with this information, the doctor is able to diagnose conditions such as coronary artery disease, or identify tissue damage associated with a heart attack (Cardiovascular Institute of the South, 2018). This can also help identify whether or not the patient will benefit from surgery, which might be applied for a variety of conditions.

There are some potential limitations to the information gained through a PET scan, however. Other conditions can cause some confusing results. Rheumatoid arthritis and tuberculosis can both interfere with the imaging of a PET scan, and lead to some potentially misleading results . Additionally, small or slow growing tumours may not show up in PET scan images, as the former may simply not show up on the scan while the latter may not absorb much of the radioactive tracer in order to display properly. Blood sugar levels can interfere with accurate scans, which is why patients are usually asked to fast beforehand. The radioactive agent key for the PET scan can decay shortly after it has been injected into the body, so the timeframe for doing the scan is relatively limited (Virtual Medical Centre, 2017). If the patient wants to accelerate the rate of the radioactive tracer's decay, drinking fluids after the test can accelerate the process.

Brain Function

Much of the function of PET scans of the brain revolve around the use of glucose. Glucose functions as the main fuel of the brain, so the PET scan works by attaching the radioactive tracer to the glucose compounds, allowing the metabolic processes of the brain to light up, relative to their usage of glucose (Krans, 2018). Therefore, much of what a PET scan does is related to seeing what parts of the brain are using glucose to different degrees of efficacy.

This method can be used to detect Alzheimer's disease, depression, head trauma, epilepsy or Parkinson's disease (Krans, 2018). With Alzheimer's, it can see which parts of the brain are using glucose at what rate, as parts of the brain that are affected by Alzheimer's tend to use glucose more slowly than other parts (Brazier, 2017). Tumours may, by contrast, may show up as "bright spots" since they tend to use glucose more aggressively than normally functioning parts of the brain (Radiologyinfo.org, 2019). The parts of the brain that epilepsy affects can be similarly found out, and it can help determine the plausibility of surgery as a treatment plan. In Alzheimer's and Parkinson's, often the PET scan allows for the chance to identify which parts of the brain are functioning differently from the other parts of the brain (Better Health, 2019).

Use in Animals

In addition to its use in human beings, there are some applications of PET scans for animals.

University of Saskatchewan PET-CT Machine

The University of Saskatchewan is one of only five places in North America, and the first and only in Canada, to have a PET-CT machine for use with animals. However, as you might expect with this sort of expensive use of equipment, it is being used in large part for academic purposes, but those academic purposes could have very positive implications for veterinary purposes, as it may allow for extensive testing of the effectiveness of varying forms of treatment for animals (Gunville, 2018).

This also means that for the animals that use the PET-CT machine, it will be possible to diagnose cancer much faster. There are a number of benefits to oncology, the study of cancer, through the information that the PET-CT machine at the U of S can provide. The scans can be used to determine the extent of a tumour, which can then later inform as to what kind of treatment is needed, just as it is with human patients.

The upshot of this is the research opportunities that are available to researchers. Much of what applies to animals also applies to human beings, and there will be potentially useful information gained about diseases common to human and non-human animals. The PET-CT machine will allow for better information in trials of animals before they are trialed in human beings. It may allow us to learn key information in how we treat and respond to cancer, for instance (Hill, 2019).

The PET-CT machine will still be used for diagnosing medical problems with sick animals. For those that can afford to have the scan done, this does seem to be certainly more convenient than having to leave the country all-together in order to get some answers as to a family pet's condition.

Risks

While the procedure is generally considered safe, there are some minimal risks associated with a PET scan. The most obvious is that you are being injected with a radioactive tracer into your body in order to allow the scan to happen. This is generally non-harmful, because the radiation level is quite low, but it does carry some risk of radiation exposure. It is possible to have an allergic reaction to the radioactive tracer, though that is rare and the effects when someone does have a reaction are generally fairly muted. Allergic reactions may occur if a person has any adverse reaction to iodine, aspartame or saccharin, since such can be found in the radioactive tracer. A reaction may be more likely for those with asthma, heart disease, dehydration or kidney disease (Krans, 2018). People who are pregnant are advised to not get a PET scan due to the possibility of the radioactive agent affecting the in utero child, and those who are breastfeeding may need to find alternative ways to feed their child (Brazier, 2017). Of course, since the radioactive agent is usually injected, those with a phobia of needles may have some difficulty as well. The test may also be difficult for those who are claustrophobic, as they will be asked to stay within an enclosed space for an extended period of time.

While we should bear specific circumstances in mind, generally PET scans are on the whole safe and carry minimal overall risk (Krans, 2018). However, if there is a combination PET-CT or PET-MRI scan, it does carry the potential risks of CT and MRI scans, respectively, since those scans are also being performed simultaneously. The CT scan uses a moderate amount of radiation, which increases in risk if done multiple times in short succession (Medical Imaging of Fredericksburg, 2019). MRI do not use radiation, but because they use magnetic fields, they are not safe for specific people, namely those who have implanted electronic devices, such as pacemakers, cochlear implants and neurostimulators, in their body that may be adversely affected by strong magnetic forces (Stanford Health Care, 2020). This does seem to be a common-sense restriction given how the procedure works—if you have any sort of computer implant in your body, you probably don't want to jump into essentially a giant magnet. You may also want to avoid a PET-MRI scan if you have any sort of metal in your body. A contrast dye with some specific possible allergens is also sometimes used, so being upfront about what sorts of allergies you might have, particularly if they include shellfish, iodine or some medications, would be worth mentioning.

Summary

When you consider the range of uses for a PET scan, and how the imaging can be extended for use through being combined with a CT or MRI scan, there is a wide range of utility for a PET scan, although it tends to be more used for diagnosing conditions that have to do with heart/cardiovascular conditions, brain/central nervous system conditions and the diagnosis of various forms of cancer. PET scans remain fairly expensive, so it is possible that this prevents it from being used in as many situations as it might otherwise be considered useful, but it is a powerful diagnostic tool at the times in which it is utilized. Fortunately, there are relatively few risks in using a PET scan, although it may not be entirely safe for those who are pregnant, and some care may need to be used for those who are breastfeeding. In addition, there are several applications to using PET scans for non-human animals, which may have some research implications for disease in human beings, as well as allowing us to care for disease and dysfunction in household pets and other animals we routinely encounter. While there are other possible medical imaging techniques, nothing quite replicates the information in a PET scan, and therefore, it has a firmly established niche in the medical science landscape of today.

What are Opposing or Alternative Imaging Technologies to PET Scans?

By Megha Sharma

SPECT

SPECT is a nuclear imaging filter that incorporates figured tomography (CT) and a radioactive tracer. The tracer is the thing that permits specialists to perceive how blood streams to tissues and organs.Before the SPECT examines, a tracer is infused into your circulatory system. The tracer is radiolabeled, which means it discharges gamma beams that can be identified by the CT scanner. The PC gathers the data radiated by the gamma beams and shows it on the CT cross-segments. These cross-segments can be added back together to frame a 3D picture of your brain (Spine,n.d) .The radioisotopes ordinarily utilized in SPECT to mark tracers are iodine-123, technetium-99m, xenon-133, thallium-201, and fluorine-18. These radioactive types of normal components will go through your body and be recognized by the scanner.

Different medications and different synthetic substances can be named with these isotopes.The kind of tracer utilized relies upon what your primary care physician needs to quantify (Spine,n.d). For instance, if your primary care physician is taking a gander at a tumor, the person may utilize radiolabeled glucose (FDG) and watch how it is processed by the tumor. The test varies from a PET output in that the tracer stays in your circulation system instead of being consumed by encompassing tissues, in this manner restricting the pictures to zones where blood streams. SPECT examines are less expensive and more promptly accessible than higher goal PET sweeps. A SPECT check is basically used to see how blood moves through conduits and veins in the cerebrum. Tests have shown that it very well may be more touchy to cerebrum injury than one or the other MRI or CT checking on the grounds that it can identify decreased blood stream to harmed destinations. This sort of checking is likewise valuable in diagnosing pressure cracks in the spine (spondylolysis), blood denied (ischemic) spaces of the cerebrum following a stroke, and tumors (Spine,n.d). An exceptionally prepared atomic medication technologist will play out the test in the Nuclear Medicine branch of the clinic, or at an outpatient imaging focus. In the first place, you will get an infusion of a limited quantity of radioactive tracer.

You'll be given some information about 10-20 minutes until the tracer arrives at your cerebrum. Then, you'll lie serenely on a scanner table while a unique camera turns around your head. Make certain to stay as still as possible, take exact pictures.Once the output is finished, you can leave. Make certain to drink a lot of liquids to flush the tracer from your body (Spine,n.d).The tracer is radioactive, which implies your body is presented to radiation. This openness is restricted, nonetheless, in light of the fact that the radioactive synthetics have short half-lives. They separate rapidly and are taken out from the body through the kidneys.The long haul hazard of radiation openness is generally worth the advantages of diagnosing genuine ailments.

Your openness hazard could differ, be that as it may, contingent upon the number of CT or different outputs you have had. In the event that you have worries about your total radiation openness, converse with your doctor.Women who are pregnant or nursing ought not go through a SPECT scan (Spine,n.d). Some individuals may have a hypersensitive response to the tracer or the differentiation specialist.

X-RAY

A X-ray is a typical imaging test that has been utilized for quite a long time. It can help your PCP see within your body without making a cut. This can help them analyze, screen, and treat numerous ailments. Various sorts of X-ray are utilized for various purposes (Medical information and health advice you can trust,n.d). For instance, your PCP may arrange a mammogram to look at your bosoms. Or then again they may arrange an X-ray with a barium douche to draw a nearer take a gander at your gastrointestinal tract.There are a few dangers implied in getting an X-ray.

However, for a great many people, the potential advantages exceed the dangers. Converse with your primary care physician to study what is appropriate for you. Your PCP may arrange an X-ray to: inspect a territory where you're encountering agony or distress, screen the movement of an analyzed sickness, for example, osteoporosis and check how well an endorsed therapy is functioning. X-beams are standard methodology. Much of the time, you will not have to find exceptional ways to plan for them (Medical information and health advice you can trust,n.d). Contingent upon the space that your PCP and radiologist are inspecting, you might need to wear free, open to garments that you can without much of a stretch move around in. They may request that you change into an emergency clinic outfit for the test.

They may likewise request that you eliminate any gems or other metallic things from your body before your X-ray is taken.Always tell your primary care physician or radiologist on the off chance that you have metal inserts from earlier medical procedures. These inserts can obstruct X-ray from going through your body and making an unmistakable image.In a few cases, you may have to take a different material or "differentiation color" before your X-ray. This is a substance that will help improve the nature of the pictures. It might contain iodine or barium compounds. Contingent upon the justification of the X-ray, the difference color might be given in an unexpected way, including: through a fluid that you swallow,injected into your body and given to you as a bowel purge before your test. In case you're having an X-ray to inspect your gastrointestinal parcel, your primary care physician may request you to quick for a specific sum from time previously (Medical information and health advice you can trust,n.d). You should abstain from eating anything while you're fasting. You may likewise have to stay away from or limit drinking certain fluids. Now and again, they may likewise request that you take prescriptions to get out your insides.

A X-ray technologist or radiologic can play out a X-ray in a clinic's radiology division, a dental specialist's office, or a facility that works in demonstrative procedures.Once you're completely arranged, your X-ray expert or radiologist will reveal to you how to situate your body to make clear pictures. They may request that you lie, sit, or remain in a few situations during the test. They may take pictures while you remain before a specific plate that contains X-ray film or sensors. Sometimes, they may likewise request that you lie or sit on a specific plate and move an enormous camera associated with a steel arm over your body to catch X-ray images.It's critical to remain still while the pictures are being taken (Medical information and health advice you can trust,n.d). This will give the most clear pictures conceivable. The test is done when your radiologist is happy with the pictures assembled. X-ray utilizes limited quantities of radiation to make pictures of your body. The degree of radiation openness is viewed as safe for most grown-ups, however not for a creating infant. In case you're pregnant or trust you could be pregnant, tell your primary care physician before you have an X-ray. They may propose an alternate imaging strategy, like a MRI. In case you're having an X-ray never really analyze or deal with an agonizing condition, like a messed up bone, you may encounter agony or uneasiness during the test. You should stand firm on your body in specific situations while the pictures are being taken. This may cause you agony or distress. Your primary care physician may suggest taking agony medication previously. In the event that you ingest a different material before your X-ray, it might cause results. These include hives,itching,nausea,lightheadedness, and a metallic desire for your mouth (Medical information and health advice you can trust,n.d). In extremely uncommon cases, the color can cause a serious response, like anaphylactic stun, exceptionally low circulatory strain, or heart failure. In the event that you speculate you're having an extreme response, contact your primary care physician right away.

After your X-beam pictures have been gathered, you can change once more into your normal garments. Contingent upon your condition, your PCP may encourage you to approach your ordinary exercises or rest while you're sitting tight for your outcomes. Your outcomes might be accessible around the same time as your strategy, or later. Your PCP will survey your X-ray and the report from the radiologist to decide how to continue. Contingent upon your outcomes, they may arrange extra tests to build up an exact determination (Medical information and health advice you can trust,n.d). For instance, they may arrange extra imaging examines, blood tests, or other symptomatic measures. They may likewise endorse a course of treatment.

MRI

Magnetic resonance imaging (MRI) is a clinical imaging strategy that utilizes an attractive field and PC produced radio waves to make nitty gritty pictures of the organs and tissues in your body.Most MRI machines are huge, tube-molded magnets (Mayo Foundation for Medical Education and Research,n.d). At the point when you lie inside a MRI machine, the attractive field incidentally realigns water particles in your body. Radio waves cause these adjusted particles to deliver faint signs, which are utilized to make cross-sectional MRI pictures — like cuts in a portion of bread.The MRI machine can likewise create 3D pictures that can be seen from various points. X-ray is a noninvasive path for your primary care physician to look at your organs, tissues and skeletal framework (Mayo Foundation for

Medical Education and Research,n.d). It creates high-goal pictures of the body that help analyze an assortment of issues. X-ray is the most as often as possible utilized imaging trial of the cerebrum and spinal rope.

It's regularly performed to help analyze: aneurysms of cerebral vessels, issues of the eye and internal ear, different sclerosis, spinal rope disorders,stroke,tumors and cerebrum injury from injury. An exceptional sort of MRI is the practical MRI of the cerebrum (fMRI). It produces pictures of blood stream to specific spaces of the mind. It tends to be utilized to inspect the cerebrum's life structures and figure out what parts of the mind are dealing with basic functions.This recognizes significant language and development control regions in the minds of individuals being considered for cerebrum medical procedure (Mayo Foundation for Medical Education and Research,n.d). Useful MRI can likewise be utilized to survey harm from a head injury or from problems, for example, Alzheimer's disease.

MRI utilizes amazing magnets, the presence of metal in your body can be a wellbeing peril whenever pulled into the magnet. Regardless of whether or not pulled into the magnet, metal articles can contort the MRI picture. Prior to having a MRI, you'll probably finish a poll that incorporates whether you have metal or electronic gadgets in your body. On the off chance that you have tattoos or lasting cosmetics, find out if they may influence your MRI. A portion of the hazier inks contain metal.Before you plan a MRI, tell your PCP on the off chance that you believe you're pregnant. The impacts of attractive fields on hatchlings aren't surely known. Your primary care physician may suggest an elective test or deferring the MRI. Additionally tell your primary care physician in case you're bosom taking care of, particularly in case you're to get contrast material during the procedure.It's likewise imperative to talk about kidney or liver issues with your PCP and the technologist, since issues with these organs may restrict the utilization of infused contrast specialists during your output (Mayo Foundation for Medical Education and Research,n.d).

Prior to a MRI test, eat ordinarily and keep on taking your typical prescriptions, except if in any case trained.

You will normally be approached to change into an outfit and to eliminate things that may influence the attractive imaging. The MRI machine resembles a long restricted cylinder that has the two closures open. You rest on a mobile table that slides into the launch of the cylinder. A technologist screens you from another room. You can converse with the individual by microphone. If you have a dread of encased spaces (claustrophobia), you may be given medication to help you feel lethargic and less restless. A great many people get past the test without difficulty. The MRI machine makes a solid attractive field around you, and radio waves are aimed at your body. The strategy is effortless. You don't feel the attractive field or radio waves, and there are no moving parts around you.

During the MRI check, the inside piece of the magnet produces monotonous tapping, pounding and different commotions. You may be given earplugs or have music playing to help block the noise (Mayo Foundation for Medical Education and Research,n.d). In a few cases, a differentiation material, commonly gadolinium, will be infused through an intravenous (IV) line into a vein in your grasp or arm. The differentiation material upgrades certain subtleties. Gadolinium infrequently causes unfavorably susceptible reactions. An MRI can last somewhere in the range of 15 minutes to over 60 minutes. You should keep still since development can obscure the subsequent images. During a useful MRI, you may be approached to play out various little errands — like tapping your thumb against your fingers, scouring a square of sandpaper or addressing straightforward inquiries. These aides pinpoint the parts of your mind that control these activities. A specialist exceptionally prepared to decipher MRIs (radiologist) will examine the pictures from your sweep and report the discoveries to your primary care physician (Mayo Foundation for Medical Education and Research,n.d). Your primary care physician will examine significant discoveries and following stages with you.

fMRI

Functional magnetic resonance imaging (fMRI) portrays changes in deoxyhemoglobin focus subsequent to task-initiated or unconstrained adjustment of neural digestion (Glover, 2011). Since its origin in 1990, this strategy has been generally utilized in a large number of investigations of insight for clinical applications like careful arranging, for observing treatment results, and as a biomarker in pharmacologic and preparing programs. Specialized improvements have tackled the vast majority of the difficulties of applying fMRI practically speaking. These difficulties incorporate low differentiation to commotion proportion of BOLD signs, picture bending, and sign dropout (Glover, 2011). All the more as of late, consideration is going to the utilization of example order and other measurable techniques to draw progressively complex surmisings about psychological cerebrum states from fMRI information.

This paper surveys the strategies, a portion of the difficulties and the eventual fate of fMRI. Utilitarian Magnetic Resonance Imaging (fMRI) is a class of imaging strategies created to exhibit territorial, time-fluctuating changes in mind digestion (Glover, 2011). These metabolic changes can be ensuing to task-instigated psychological state changes or the consequence of unregulated cycles in the resting mind. Since its beginning in 1990, fMRI has been utilized in an astoundingly huge number of studies in intellectual neurosciences, clinical psychiatry/brain science, and presurgical arranging (somewhere in the range of 100,000 and 250,000 passages in PubMed, contingent upon watchwords). The ubiquity of fMRI gets from its far and wide accessibility (can be performed on a clinical 1.5T scanner), non-obtrusive nature (doesn't need infusion of a radioisotope or other pharmacologic specialist), moderately minimal effort, and great spatial goal. Progressively, fMRI is being utilized as a biomarker for illness, to screen treatment, or for contemplating pharmacological adequacy (Glover, 2011). Subsequently, it is important to audit the fMRI contrast components, the qualities and shortcomings, and transformative patterns of this significant instrument.

fMRI is obviously founded on MRI, which thusly utilizes Nuclear Magnetic Resonance combined with angles in attractive field to make pictures that can join various kinds of differentiation, for example, T1 weighting, T2 weighting, weakness, stream, etc.7 In request to comprehend the specific difference instrument transcendently utilized in fMRI it is important to initially talk about mind digestion. Every one of the cycles of neural motioning in the cerebrum, including development and spread of activity possibilities, restricting of vesicles to the pre-synaptic intersection, the arrival of synapses across the synaptic hole, their gathering and recovery of activity possibilities in the postsynaptic structures, searching of overabundance synapses, and so forth, require energy as adenosine triphosphate (ATP).

This nucleotide is created mainly by the mitochondria from glycolytic oxygenation of glucose, and its creation brings about carbon dioxide as a side-effect (Glover, 2011). At the point when a district of the cerebrum is up-directed (for example actuated) by an intellectual undertaking, for example, finger tapping, the extra neural terminating and other expanded flagging cycles bring about a privately expanded energy necessity, thus coming about in up-managed cerebral metabolic pace of oxygen (CMRO2) in the influenced mind locale. As the neighborhood stores of oxygen in tissues adjoining vessels are momentarily devoured by glycolysis and byproducts develop, different synthetic signs (CO_2, NO, $H+$) cause a vasomotor response in blood vessel sphincters upstream of the hairlike bed, causes enlargement of these vessels. The expanded blood stream acts to reestablish the neighborhood [O2] level needed to defeat the transient shortfall; nonetheless, for reasons that are as yet not completely seen, more oxygen is conveyed than is expected to counterbalance the expansion in CMRO2 (Glover, 2011). Accordingly, neural up-guideline results at first in a development of deoxygenated hemoglobin [Hb] and an abatement in deoxygenated hemoglobin [HbO2] in the intra-and extravascular spaces, followed inside a little while by a vasodilatory reaction that turns around the circumstance to bring about an expansion in [HbO2] and lessening in [Hb] over that in the resting condition.

This arrangement of cycles is portrayed as the hemodynamic reaction to the neural occasion. Errand initiation fMRI contemplates look to incite changed neural states in the mind as the visual, hear-able or other boost is controlled during the output, and enactment maps are acquired by contrasting the signs recorded during the various states (Glover, 2011). Accordingly, it is essential to gather each picture in a depiction mode to stay away from head movement and physiological cycles of breath and cardiovascular capacities from infusing commotion signals inconsequential to the neural handling being cross examined.

By and large, most fMRI is performed utilizing an Echo Planar Imaging (EPI) technique, which can gather information for a two dimensional picture in around 60 ms at common goals. Normally, entire cerebrum checks with ~ 32 2D cuts are gained with a redundancy time (TR) of 2s/volume (Glover, 2011). Each voxel in the subsequent sweep delivers a period arrangement that is consequently examined in understanding to the errand plan. The commonplace fMRI task initiation test uses visual, hear-able or different upgrades to then again actuate at least two diverse psychological states in the subject, while gathering MRI volumes constantly as portrayed previously. With a two-condition plan, one state is known as the trial condition, while the other signifies the control condition, and the objective is to test the theory that the signs vary between the two states. Utilizing a square plan, the preliminaries are masterminded to switch back and forth between the trial and control conditions, as demonstrated in, with each square commonly being two or three many seconds long. The square plan is ideal for recognizing enactment, however a jittered occasion related (ER) plan is predominant when portrayal of the adequacy or timing of the hemodynamic reaction is wanted.

In the ER configuration, task occasions are moderately concise and happen at non-consistent intervals between preliminary spans with longer times of control condition, which permits the hemodynamic reaction to return all the more completely to benchmark (Glover, 2011). Jittering the circumstance serves to test the hemodynamic reaction with higher transient recurrence in the general time arrangement, however may likewise be utilized to instigate an ideal psychological procedure, e.g., to keep away from an expectant reaction or look after consideration. Generally, the MRI physical science and innovation advancement behind BOLD fMRI acquisitions are developed, and the tradeoffs between procurement speed, goal, SNR, signal dropout and difference are surely known. Throughout the long term, various examiners have endeavored to create options in contrast to BOLD difference utilizing direct neural current detection9, despite the fact that at this point it is perceived that the feeble size of the neural current sign comparative with physiological commotion makes an advancement improbable. Another option is the utilization of dissemination weighted imaging to exhibit enactment related changes in populaces of bound versus free water dissemination (Glover, 2011). A potential benefit is that such dispersion related changes may have more fast reactions than BOLD strategies.

In any case, again the signs are more fragile than BOLD differentiation and their biophysical beginning is as yet indistinct. Different tests have revealed the utilization of twist reverberation instead of inclination reverberation acquisitions of BOLD differentiation, particularly at higher fields where T2* is foreshortened. While an unassuming examination exertion will proceed in improving obtaining innovation, the main part of exploration in the advancement of fMRI has moved to its application to addressing more perplexing inquiries in intellectual neuroscience (Glover, 2011). One promising territory is that of utilizing actuation maps as contribution to grouping and state change calculations to foresee or order psychological conduct, for example, anticipating cerebrum states. Other arising employments of fMRI incorporate the improvement of quantitative measures, for example biomarkers for infection or checking conduct alteration like understanding issues.

A preventative note, in any case, is that due to the little BOLD reactions average of intellectual cycles, most examinations are restricted to utilizing bunch measurements to make surmisings about populaces instead of about people. Subsequently fMRI's utilization in measuring singular attributes may keep on being restricted to those undertakings for which generally solid BOLD reactions are noticed, like essential tangible frameworks (Glover, 2011). Resting state organizations and their alteration by infection conditions like Alzheimer's, melancholy and other mental disorders32 are acquiring consideration. Nonetheless, there is developing mindfulness that these organizations might be significantly more intricate in their spatio-transient elements than beforehand thought17, and considerably more work is demonstrated to comprehend their job and utility in foreseeing singular conduct/physiology (Glover, 2011).

At long last, input from constant fMRI has appeared to permit subjects to learn torment decrease procedures, upgrade sensorimotor control and to control significant cerebrum locales in mind-set problem tests. The peruser is likewise alluded to Bandettini for extra contemplations in regards to the eventual fate of fMRI. Utilitarian MRI has delighted in an energizing advancement course with a remarkable development in distributed examinations since its commencement in the mid 90's, and it has gotten typical for clinical uses, for example, presurgical arranging, major intellectual neuroscience examinations, conduct change and preparation (Glover, 2011). Educated by fMRI, more modern displaying of mind networks is sure to prompt new degrees of comprehension of the human cerebrum.

What Misinformation or Conspiracy Theories Exist Regarding PET Scans?

By Rosalie Sullivan

Misinformation

Misinformation is information that is distorted, inaccurate, false, misleading, and deceitful. It is spread around with the intention of being dishonest. The information is created and shared with the intention of lying to others. It is crafted in a misleading and distorted way, and it is designed to trick and deceive other people. There is usually a purpose in the creation of misinformation, but not always.

Conspiracy Theories

Conspiracy theories, similarly, are also crafted in a way that can mislead other people. A conspiracy theory is a theory that suggests that what we believe to be is true, is actually false. The theory explains that we are all being lied to, being deceived, by a higher authority. Conspiracy theories can involve a variety of topics, including medical topics. Essentially, it is a theory that asserts that we are being lied to, or that something is being covered up.

PET Scans and Conspiracy Theories

PET Scans, also known as positron emission tomography, require the use of radioactive tracers, and because of this, a wide variety of conspiracy theories and myths surround the body imaging technology. Radiation is a material that draws a lot of attention onto itself, and there are a lot of conspiracy theories surrounding the radioactive material. Many people believe that radiation, in every instance, is a harmful and destructive energy. This, of course, is not the case. Radiation is not always harmful to the human body, and the radioactive material used in PET scans is essentially harmless. The radioactive tracers, used in PET scans, are given off in very small doses, so there is less risk to the patient. The risk of negative side effects is also astoundingly low, as the injection of radioactive glucose is minimal (Portnow et al., 2013). Receiving a PET scan is not a risky procedure, and the conspiracy theories surrounding them are just that—conspiracy theories.

Radiation and its Negative Image in the Media

Yet despite the fact that the procedure is a low risk operation, many people spread misinformation and conspiracy theories about positron emission tomography. This is because PET scans require the use of radioactive tracers, and radiation has a very negative image in the media. When the general public thinks about radiation, they think about mutations, cancer, hazardous material, and other negative commercialized concepts of radiation. Sometimes people will think about the world wide disaster that was the Chernobyl nuclear accident. Even today, Chernobyl remains a ghost town in the Ukraine, as dangerous levels of radiation continue to circulate throughout the abandoned town. Almost thirty four years later after the initial accident, Chernobyl is still a radioactive hotspot, and people are still avoiding the town (Alimov, 2020). Radiation can have long term devastating consequences, and people are well aware of this.

Radiation is something that has a negative image in the media, and most people perceive it as a detrimental tool. Due to the negative image of radiation, many people assume that PET scans are dangerous because they use radioactive tracers. People spread misinformation about PET scans, claiming that they can make one sick and even possibly give them cancer. Since the PET scans use radiation in their treatment, they must be dangerous, right? Radiation decimated the Chernobyl area and ecosystem, and the "consequences of [the] Chernobyl [accident]" (Alimov, 2020) are still present today in the Ukraine. This must mean that consuming any amount of radiation is detrimental to your health, and that by undergoing a PET scan—you are putting yourself at risk. This, however, is not the case.

PET scans can be an efficient medical procedure when it comes to scanning the human body, and as previously stated, the risk to the patient is minimal. Even though positron emission tomography uses radiation in its medical procedure, the doses of radioactive glucose are small and the risk of negative side effects is low. One cannot significantly worsen their health or give themselves cancer by undergoing a PET scan. PET scans are an excellent way to examine and evaluate brain activity, and one will not be putting themselves at risk if they undergo the medical procedure (Portnow et al., 2013). The only major risk involved in a PET scan is the potential danger of undergoing a major allergic reaction (Mayo Clinic, 2021). Allergic reactions, however, are treatable and can be dealt with easily and safely. The probability of a major allergic reaction taking place is immensely low, and allergic reactions only occur in very rare instances (Mayo Clinic, 2021). If you undergo a PET Scan, you will most likely not experience a major allergic reaction and you will be perfectly fine. The radiation used in positron emission tomography is essentially harmless, and there is no risk to the patient.

Yet despite this, people continue to spread misinformation and conspiracy theories about the body imaging technology—claiming that is both harmful and dangerous. Radiation's negative image in the media continues to flourish and grow, and more and more people are growing suspicious of the safe medical procedure. Since radiation can cause major disasters such as the Chernobyl incident, radiation 'must' be a toxic and dangerous material. Conspiracy theories about positron emission tomography and the radiation it uses, will continue to circulate throughout the media for years to come. As long as a negative image of radiation is promoted, people will spread misinformation about PET scans—claiming that its radioactive material is deadly and unsafe.

PET Scan Conspiracy Theories Involving High Costs

Another conspiracy theory that surrounds positron emission tomography involves its high cost. Receiving a PET scan can be very expensive, and many people are irritated by these high costs. Why should someone pay thousands of dollars to get injected with radiation, a 'negative' substance? For many people, the high cost of a PET scan is completely unreasonable and unacceptable. People do not want to pay thousands upon thousands of dollars for radioactive glucose to be injected into their body—they do not see it as a worthy health investment.

Due to this, misinformation and conspiracy theories about PET scans have begun to spread across the media. There is this idea that PET scans are a scam invented to steal money from its patients. People will pay high costs in order to undergo treatment, and the radioactive tracers will just worsen the patient's condition—forcing them to spend even more money on more expensive treatment. Essentially, the conspiracy theories claim that the radioactive glucose used in positron emission tomography will just worsen the patient's health, the radiation will worsen their cancer. Once their condition is worsened, the patient will have to spend even more money on more treatment—treatment that is all expensive. The conspiracy theories claim that positron emission tomography is just one big scam where a patient continually pays for treatment that, in reality, degrades their health. This, of course, is not true, and these conspiracy theories are horrendously wrong. They misrepresent what PET scans are, why they cost so much, and many other factors that surround the body imaging technology. Even though Pet scans can be expensive, they are a worthwhile medical investment if you need to evaluate your body's condition.

Recently, however, the high costs of PET scans have begun to change. In the long term, PET scans can now actually be a cheaper alternative to other medical expenses (Rohren, 2014). In recent years as well, "the cost of the scanners [and PET scanning] has declined" (Rohren., 2014), and receiving a PET scan is at the cheapest it has ever been. Going forward in the future, it is expected that positron emission tomography's high costs will continue to decline and become more affordable. Hopefully when PET scans are significantly cheaper, the misinformation and conspiracy theories surrounding them and their high costs will begin to diminish. PET scans are not a scam after all, and they are a very important medical procedure. The body imaging technology used in a PET scan has "proven to be an invaluable tool in the diagnosis, staging, and management of [an] oncologic patient" (Rohren, 2014), and it is a technology that we must continue to utilize in the future. Positron emission tomography is not a scam created with the intention of stealing money from its patients—it is a piece of medical technology that will continue to grow important in coming years. All of the conspiracy theories surrounding the high costs of PET scans are inaccurate and overly paranoid.

The Common Misconception Surrounding FDG Uptake

Another common misconception that surrounds positron emission tomography technology involves the discussion of FDG uptake. Many people believe that on "a pet scan, anything with FDG uptake is abnormal" (docpanel, 2020). This, of course, is not the case. Having a high FDG uptake can be a cause for concern, but it is not always negative. Possessing a high FDG uptake can be a sign of cancer, but just because it is there—does not mean the patient has cancer. A high FDG uptake can appear for a variety of reasons and determining what the FDG uptake is indicative of, can be a complicated process (docpanel, 2020). To "understand all the variables and intricacies" (docpanel, 2020) involved in FDG PET scans, one must have a lot of experience and specialization in the medical field. Since a high FDG uptake is not always a determinant of whether or not someone actually has cancer, it "can cause unnecessary alarm and concern" (docpanel, 2020) for many patients. This common misconception tricks patients into believing that they have cancer, that they could be dying, when that is not necessarily the case. A high FDG uptake can often mean that a cancer is developing in the patient, but that is not always the case—and it is important to keep that in mind.

This medical misconception causes a lot of emotional damage in its patients, and it can cause them a lot of grief. Eliminating this misconception is vital in ensuring that the patients can receive their results with a clear head. If a common misconception tricks them into thinking that they have cancer, that they could be dying, they will most likely experience unnecessary pain. Going forward in the future, misconceptions and conspiracy theories, such as this one, need to be eliminated to ensure patient satisfaction. Patients should not have to overly stress over their results; they should be granted peace of mind when undergoing a difficult medical procedure.

Positive Misinformation Surrounding PET Scans

There are, however, also positive myths and misinformation surrounding PET scans. Many people believe that PET scans are the most efficient way to determine whether or not something in the human body is malignant (docpanel, 2020). This, however, is not the case. PET scans are not very good at detecting viruses, infections, or malignant cancers, in the human body. They lack the proper scanning technology to do so, and they are more adept at examining and identifying cancer that has already been discovered. PET scans struggle with detecting malignancy in the body, but they are capable of staging even the smallest bits of diagnosed cancer (docpanel, 2020). Yet despite the fact that PET scans are more adept at staging cancers in the human body, rather than discovering them, many people believe that PET scans are fully capable of detecting developing malignancy. This is a common misconception that exists in the modern world, but it is a positive misconception. People think that PET scans are more efficient than they actually are, and they give them too much credit. This misconception does spread false information, but the misinformation is not as harmful as previously discussed misconceptions and conspiracy theories.

There are, however, still consequences to these misconceptions. One may find themselves disappointed with their PET scan, if the medical procedure was not able to offer everything the patient wanted. If the patient believed that receiving a PET scan would offer them more insight on a virus, infection, or malignant cancer, flourishing in their body, they would be sorely disappointed. With all of the misconceptions and conspiracy theories that float around our modern world, on the internet, we must make sure to properly educate ourselves on any procedure we might go through. Nobody wants to be disappointed.

Misconceptions and conspiracy theories about PET scans, even the positive ones, need to be snuffed out. They offer no real insight into positron emission tomography telenology, and they just further complicate our already complicated world. Conspiracy theories are unneeded in the medical field, and misconceptions can be dangerous. If one does not understand the science behind a treatment they are going through, they may be putting themselves at risk. Going forward in the future, we need to ensure that people who are receiving PET scans are properly educated on what a PET scan entails. We need to make sure that they have not gotten any of their information from common misconceptions, myths, and conspiracy theories.

Conclusion

In conclusion, many misconceptions, myths, and conspiracy theories, surround PET scanning technology for a variety of reasons. PET scans require the use of radioactive tracers, and radiation has a negative image in the media. When one thinks about radiation and its effect on the world, they think about world wide disasters such as the Chernobyl incident. Radiation is seen as a negative energy in the world, and many people do not want to consume any part of it. Since PET scans require the patient to consume radioactive glucose, however, many are reluctant to undergo this treatment. Another conspiracy theory that surrounds PET scans involves its high cost. The conspiracy theory claims that PET scans are, essentially, a scam created in order to steal patient's money. This, of course, is not the case. PET scans are actually becoming more and more affordable in this modern age of technology, and the high price of PET scans is expected to drop. Another common misconception that involves PET scans has to do with FDG uptake. Many people believe that having a high FDG uptake is a cause for concern, and many people believe that it means that the patient has cancer. This, however, is not actually true. Just because a patient experiences an FDG uptake, does not mean that they have cancer. This misconception causes unnecessary problems, emotional harm, and worry in patients. Lastly, however, there are also positive myths and misconceptions surrounding PET scans. Many people believe that PET scans are the best way to determine whether or not something is malignant in the human body, when that is not the case. PET scans are not adept at determining whether or not something is malignant, and it is better at staging cancer.

All of these misconceptions and conspiracy theories, however, cause considerable harm to both patients and the medical field. When one is receiving treatment, they must be educated about the treatment they are receiving. Going forward in the future, people need to make sure that what they are learning about PET scans is actually true, and the medical field needs to ensure that their patients are knowledgeable about the treatment they are receiving.

How are PET Scans Talked About Commonly and in Popular Culture?

By Katerina Bavaro

The confusion between PET scans and other radiological interventions

PET scans are often confused with CT scans though they are commonly utilized together as an intervention of diagnostic radiology. PET scans as previously mentioned are a type of medical imaging that helps reveal how organs and tissues are functioning (Mayo Clinic, 2020). CT scans combine X-ray images into a series taken from different angles around the body and use computer processing to create cross-sectional images, or slices of the bones, blood vessels, and soft tissues of the body (Mayo Clinic, 2020). Due to the sheer amount of technologies involved in the medical imaging field and the fact that there are always new discoveries and revisions of previous procedures, it is difficult for the general public to stay informed. Later in the chapter, resources will be provided in order to further the understanding of current medical imaging procedures and their functions. To avoid further confusion, listed below is a table of the common types of diagnostic radiology which explains what the technique is, what is being done, and the advantages and disadvantages of it.

Table 1: Common types of diagnostic radiology exams

Technique	What is being done	Advantages	Disadvantages
Computed tomography (CT)	Combines x-ray images taken from angles around the body to create cross-sectional images Fluoroscopy	(Fernandez- Friera et al., 2012) • High spatial resolution • Absorption proportional to the concentration of contrast	(Fernandez- Friera et al., 2012) • Low sensitivity to molecular imaging • Susceptibility to motion artifacts
Fluoroscopy	Continuous x-ray beam passed through the body part to be examined. Live moving pictures.	• Physician can see a live image of the body's internal organs • Dynamic and functional information available	• May display overlapping anatomy • Poor soft tissue resolution • Ionizing radiation
Magnetic resonance imaging (MRI)	Uses a magnetic field and computer-generated radio waves in order to create images of organs and tissues.	(Fernandez- Friera et al., 2012) • High spatial resolution • Availability of equipment • No radiation	(Fernandez- Friera et al., 2012) • Low sensitivity to molecular imaging • Susceptibility to motion artifacts
Nuclear Medicine (bone scan, etc.)	Uses radioactive materials called radiotracers in order to create an image of the skeleton	• Early detection of bone cancer • Early detection of cancer which has spread from other parts of the body • Can detect other bone abnormalities	• Radiation
X-rays	X-ray beams pass through the body and are absorbed depending on density of material passed through.	• Easy to diagnose • Medium image quality	• Ionizing radiation
Positron emission tomography (PET)	Is used to see how organs and tissues are working. Radiotracer collects in areas of high chemical activities • Also can be used to measure blood flow and oxygen use	(Fernandez- Friera et al., 2012) • Algorithm for attenuation (mechanism that removes soft tissue artifacts from images) • High sensitivity • High chemical affinity for molecular targets	Fernandez- Friera et al., 2012) • Short life of radiotracers • Expensive equipment • Local cyclotron required • Advanced radiochemistry

Ultrasound	Uses high-frequency sound waves to produce images of structures	(Fernandez- Friera et al., 2012) • Universally available • Inexpensive • No radiation	(Fernandez- Friera et al., 2012) • Limited penetration • Absence of molecular probes
Mammography	X-ray imaging used to examine the breast	• Early detection of breast cancer • Little radiation	• Possible overdiagnosis of breast cancer

Not mentioned in the table is functional magnetic resonance imaging (fMRI) which is a less common exam that is often confused with PET scans. It is a type of MRI that measures brain activity by looking at changes in blood flow. This is because cerebral blood flow is accompanied by neuron activity therefore when an area of the brain is activated and/or being used, blood flow increases in that specific area. Low blood flow to the brain indicates that neurons will not get the nutrients necessary which thereby can prevent them from functioning. If this happens for long enough, this can lead to adverse consequences such as a stroke.

Since an fMRI can detect these blood flow changes, it is a useful exam to detect abnormalities in the brain, evaluate the effects of stroke and other diseases, or to guide brain treatment (Radiologyinfo.org, 2019). An advantage is that just like a regular MRI, it does not use radiation. It instead uses a magnetic field and computer-generated radio waves in order to create images of organs and tissues. It also has moderately good spatial resolution.

Disadvantages would be the temporal response of the blood supply which is the basis for the exam, is poor relative to the electrical signals that define communication between neurons (The University of Edinburgh, 2019). It is also quite expensive, costing as much as $555 per hour though the slot typically takes 1.5 hours which costs $832.50 (Yale Medicine, 2020).

The story of C.Edmund Kells

There tends to be stigma surrounding PET scans as it belongs to the field of radiology also called radiation sciences. This comes from the misinformation and conspiracy theories that have formed over time in popular culture such as the ones involving the high costs of PET scans and the efficacy of FDG uptake (see previous chapter). Therefore, there is generally a negative image present among the general public. One of the most well-known cases is that of C.Edmund Kells, who was the first dentist in the United States to take X-rays of a living patient. His experiments, with such X-rays caused lesions on his hand which overtime led to the amputation of his entire left arm. The cancer then metastasized to other parts of his body, and due to the pain he was in resulted in him committing suicide.

The experiment was done in 1896, one year after Wilhelm Conrad Roentgen made the discovery of the X-ray. Kells noticed these lesions twelve years after he began his experiments, and for twenty years he went on to have various treatments and then passed in 1928 (LSU Health Sciences, 2012). This is one tragic case that occurred, however, it did contribute significantly to the field of radiology and has been seen as a major source of innovation. It has applied to other fields such as dentistry which C. Edmund Kells was a part of. It is important to remember that this case happened around one hundred and twenty five years ago and it has been ninety-three years since his death in 1928.

Many of the risks and side effects that were faced in Kells' time have been taken care of.

These include the chance of experiencing side effects from artificial radiation such as diagnostic imaging. Measured in millisievert (msv) or microsievert (µSv), one chest X-ray has approximately 0.2 msv of radiation. This can be compared to the exposure from natural sources which will be mentioned later in the chapter that is around 2.4 msv per year (IAEA,2014). For context, it was estimated that Kells had a consistent absorbed dose of 3000 rads (Sansare et al. 2011). This amount is equivalent to 30 siervets (sv) which is around 300 millisievert (msv). (Carr, 2015).

The body can absorb up to 200 rads without fatality however as the whole-body dose approaches 450 rads the death rate will approximate 50% and greater than 600 rads will almost certainly be fatal (NORD, 2015). As with the common association between radiation and cancer the risk of death from cancer in the United States has dropped by 29% from 1991 to 2017, and by 2.2% from 2016-2017 making it the largest drop in a single year (Simon, 2020).

Other precautions have also been taken in the imaging facility itself such as lead aprons or vests being used by radiologists which are also available for patients to use. Lead thyroid collars, lead gloves, and safety goggles are also used as Personal Protective Equipment (PPE). Time, Distance, and Shielding are important when it comes to ionizing radiation in order to minimize exposure time in areas with elevated radiation levels, maximize distance from the source of radiation, and use shielding to reduce and eliminate the dose of radiation. Specifically in terms of distance, for gamma rays and x-rays the radiation intensity is proportional to the square of the distance of the source which means increasing the distance by a factor of two will decrease the rate of the dose by a factor of four (United States Department of Labour, 2014).

A major concept underlying radiation protection programs is called ALARA which stands for As Low as Reasonably achievable (United States Department of Labour, 2014). This involves implementing certain measures in order to ensure occupationally that radiation dose limits are as low as possible such as the ones mentioned in this section.

Lastly, additional steps are recommended by physicians such as avoiding exercising the day before a PET-CT scan, following a special diet 12-24 hours before the scan, and only drinking water 6 hours before the scan. This is accompanied by drinking lots of water after the scan to help wash away the radioactive substance and dye out of the body (Goodman,2019). These have been found overtime to substantially decrease the risk of harm and to reduce the side effects of radiation from diagnostic imaging.

Though progress has certainly been made it does not mean that all of the issues have been solved. For instance, 80 million CT scans have been performed in the United States compared with three million done in 1980 (Harvard, 2010). These have essentially replaced complex and potentially harmful surgeries, as the benefits outweigh the negatives. Though many are still afraid of the associated cancer risks as ionizing radiation can damage DNA. Although most of the time cells can repair the damage done- sometimes it cannot which results in these DNA mutations that can contribute to cancer in the future (Harvard, 2010). Even though, as previously mentioned, X-rays only is about 0.2 msv of radiation, PET scans are around 25 msv, and CT scans are anywhere between 1 to 10 msv depending on the dose and the part of the body the exam is focusing on.

Recall that 200 rads is how much that body can absorb without fatality and when it starts to reach about 450 rads is when it starts to become concerning (NORD, 2015).

Below this section is the math explained. Specifically this example will involve 200 rads as this is the amount that the body can absorb without fatality and compare this to the amounts used during X-rays, PET, and CT scans.

Table 2: Millisievert and milligray as measures of radiation dose and exposure (Carr, 2015).

SI units	Historical dosimetry
1 Gray	100 R
1 Sievert (sv)	100 rad => 100 rem
10 mGy	1 Roentgen
10 mSV	1 rad => 1 rem

10 msv = 1 rad (multiply by 200 on each side)
10 msv x 200 = 1 rad x 200
2000 msv= 200 rads

Now, convert msv into sievert (sv) as larger numbers can make it more confusing.

2000 msv =200 rads ->   1,000 msv = 1 sv = 100 rad
Therefore:
2,000 msv = 2sv = 200 rads

Lastly let us compare this according to the approximate dosage used in diagnostic imaging.

2,000 msv/ 0.2 msv = X-rays
2,000 msv/25 msv = PET scans
2,000 msv/ 10 msv = CT scans (maximum amount)

Therefore it can be seen that the amount given to patients is significantly smaller than the amount necessary for maximum absorption. Let alone the number that is concerning in terms of side effects.

Humans are exposed to radiation from natural sources such as the sun, and radon which is a radioactive gas - it is a product of the breakdown of uranium in soil, rock, water, and building materials. However, the exposure to radiation, however, has grown from 15% in the early 1980's to 50% in 2010 in which medical imaging seems to be the main contributor (Harvard, 2010). The radiopharmaceutical that is given in order to perform a PET scan contains a radionuclide, fluorodeoxyglucose (FDG) being the most common one, and the drug itself which leads the radiopharmaceutical to the correct location (Alexander et al. 2018). The radionuclide contains a short half-life but remains traceable in the body ("Radiation in Biology", 2020). PET scans also involve gamma rays as once the radiopharmaceutical which contains the radionuclide is administered through an IV, the PET scanner is moved over the part of the body to be examined.

Once this happens positrons are emitted by the breakdown of the radionuclide. Gamma rays are created during the emission of these positrons and the scanner detects these rays (Johns Hopkins Medicine, 2021). Gamma rays remove electrons from atoms therefore they are a form of ionizing radiation which has a short wave-length, high frequency, and high energy. As a result of this high energy and thus high penetration, they are biologically hazardous. See chapter 6 for information on the specific science behind PET scans.

Common fears involving PET scans

Diagram 3.1

Here in diagram 3.1 can be seen a comparison between General diagnostic fears and fears surrounding specifically diagnostic imaging. For General diagnostic fears, one of the main issues is cost, as general practitioner visits (GP) tend to not be covered by insurance in places such as the United States. Diagnostic imaging however, will be covered if ordered by a physician and deemed medically necessary. Unhealthy habits may also be exposed, as in general checkups they measure blood pressure, height, and weight among other things which may expose risk for a variety of conditions. Lastly, part of a general checkup is having an open dialogue with a physician. Mentioning a symptom that one may have had for a while may aid in diagnosing a potentially serious condition therefore fear could be instilled.

As for fears surrounding diagnostic imaging, radiation is a large one due to the negative stigma that exists as previously mentioned. The chance of experiencing the extreme side effects of radiation after getting diagnostic imaging done is quite low, nevertheless, it is still a concern. Claustrophobia is another fear that is unique to diagnostic imaging procedures. Machines such as the PET-CT scanner can seem daunting to some, as parts of the body need to be enclosed or semi-enclosed in order for the procedure to be done. Lastly, cancer is the group of diseases people fear the most and attribute to diagnostic imaging. PET scans can be used to detect cancer ,however, there is also the small chance that it could lead to cancer in the future. Once again, the risk is relatively low and the benefits outweigh the harms done.

There is some general overlap between General diagnostic fears and fears surrounding diagnostic imaging, such as general anxiety that comes with being in a healthcare setting, and test results. As seen in the diagram,there are some obvious differences such as the ones attributable to solely diagnostic imaging.

Media depictions of PET scans and other radiological interventions

PET scans are often mentioned in medical shows such as Grey's Anatomy or House M.D due to the commonality of the medical imaging process. They have also been mentioned in other shows such as Law & Order and in movies such as The Fault in Our Stars. Generally, they tend to depict PET scans as being a negative and scary experience further contributing to the stigma surrounding such radiological interventions. Also, the association between PET scans and cancer tends to be made.

While it is true that PET scans are commonly used to detect cancer, it is not its sole function. Fluorodeoxyglucose (FDG) which is the main radiotracer used in PET scans can also find lesions that represent infection, inflammation, autoimmune processes, sarcoidosis and benign tumors (Safaie et al.). It is used because structurally it is a similar molecule to glucose that is a marker for metabolic activity. Areas where glucose consumption is high include the kidneys, brain, and cancer cells throughout the body (Alexander et al., 2018). Some of the most common uses for a PET scan, besides cancer include to help the diagnosis of cardiac conditions and brain/central nervous system disorders such as depression, epilepsy, and Alzheimer's disease. Therefore, attention has to be paid, as PET scans do not immediately diagnose cancer, they are used to show areas of abnormal uptake of the radiotracer which will show up as bright spots on the scans due to the high metabolic rate which signify cancer cells (Mayo clinic, 2020). Other diseases will show up as "hot areas" or "hot spots" on the scans.

Lastly, while PET scans are fairly accurate at detecting cancers, not all types will be visible on one. Typically ones with low glycolytic activity (use of glucose) which therefore lower metabolize and thus are not shown as bright spots are not seen. These include low grade lymphomas, bronchoalveolar carcinomas, and carcinoid tumors (Chang et al.). PET scans are usually just one part of a series of tests done in order to determine a diagnosis, they are good at staging cancers.

Some of the most extreme depictions of radiation in popular culture come from the genre of horror. Some of the early adaptations of the effects of radiation come from movies in the 1950s such as Godzilla (1954), Them! (1954), It came from beneath the sea (1955), Attack of the crab monsters (1957), Beginning of the end (1957), and the Incredible shrinking man (1957). Godzilla most famously tells the story of a monster that was regenerated by repeated nuclear tests. Japan, the country of origin for the story, was skeptical of nuclear power and the depiction of this through cinema represents this fear (Snyder,

2014). As for the other movies, they also told the stories of fictional characters created through radiological experimentation.

PET scans and the field of Radiology generally has had a unique scientific history. The global radiology information system market size was estimated at USD 730.1 million in 2020 and the revenue forecast in 2026 is expected to be USD 1.15 billion (Grandview Research, 2019). Today, PET scans are common practice and are available in almost every major hospital in WEIRD societies which stands for (Western, Educated, Industrialized, Rich, and Democractic). In some places, such as the province of Ontario located in Canada, it is covered by OHIP which is the local health insurance plan. In Canada, there are forty-five publicly funded PET scanners in operation along with seven which are privately funded (CADTH, 2015). In the United States, PET scans can be covered if a doctor orders the test and is medically necessary. If coverage is not given, the average cost is around $5,750 (Poslusny, 2018).

Though the potential of early response detection with PET/CT scans helps avoid drugs that could be possibly ineffective and expensive, it is still expensive to get the test done. Some patients require more than one scan to be done in a single year, however according to the CMS (Centers for Medicare and Medicaid services) in the United States, funding is only available for three rounds to be done per year (Orenstein, 2015). However, it is expected that in the long term, PET scans will become cheaper as new developments are being made such as the discovery that proton beams can be produced by blasting a solid target with a bright laser beam. The beam then converts some material into a plasma, a gas of electrically charged particles including protons. It is made from a titanium sapphire laser that generates a quick string of light pulses. It is calculated that firing the laser at aluminum foil produces a beam of fast-moving protons that aimed at a target material for thirty minutes is intense enough to generate the appropriate quantity of isotopes needed for a single dose for a patient in PET scanning (Ball, 2003).

Currently, in other countries around the world, there are large global inequities in PET scans, specifically PET-CT scans for cancer management. The population served by 1 PET-CT scanner in high income countries is 601 000, 3 484 000 in upper-middle income countries, 10 000 000 in lower-middle-income countries, and 166 667 000 in low-income countries (Gallach et al, 2020).

Resources to avoid misinformation

This book has provided a general overview of what PET scans are and how they are used in the field of radiology. There has also been some general advice provided (see chapter 4) on how to prepare for a PET scan by describing the necessary information needed to be disclosed to a physician as well as a thorough description of what to expect. Several additional resources are available in order for patients to gain a deeper understanding of current medical imaging procedures and their functions. These go beyond what has been mentioned in this book.

The Mayo clinic which has been cited several times in this chapter is a great resource for patients to have a general overview of certain procedures. Under Patient Care & Health Information on their website, they have a section dedicated to tests and procedures where an overview is given; why it is done, risks, how to prepare, what to expect, results, and clinical trials. Another resource is Radiologyinfo.org which is the Radiological Society of North America's patient resources website which is similar to the Mayo clinic though is specialized specifically to radiology. It is important to note that for these two websites ,they are American and thereby use American radiological procedures and practices. These are not the same as in other places of the world, such as Canada, thus these websites must be used in conjunction with communication with a physician. Lastly, a document by the WHO may be useful for physicians as well as patients. A chapter titled the risk-benefit dialogue details the conversation that should be happening between a radiologist and their patient (s). It is divided into three sections, the first being practical tips to support the risk-benefit discussion, the second being ethical issues, and lastly the different scenarios and key players involved when creating dialogue in a medical setting.

Conclusion

To conclude, the general dialogue surrounding PET scans tends to be negative in tone. This is due to the misinformation and conspiracy theories that are present. PET scans are often confused with CT scans and other common types of diagnostic radiology exams. A table has been provided in this chapter in order to aid the understanding of the differences between the techniques and the advantages and disadvantages though the benefits do outweigh the harms. The story of C.Edmund Kells, as the first person to apply x-rays to the field of dentistry, is meant to show how radiology and/or radiation sciences came to be understood and how it has improved since 1896. Natural sources of radiation such as the sun and radon which is commonly found in soil and rock are relatively harmless when it comes to their effects on health overtime. However, this combined with the uptake of CT scans being done has increased the overall amount of radiation exposure significantly from 15 to 50% as previously mentioned.

Depictions of the dangers of radiation have been around since the 1950's and have set the precedent for negative stigma. PET scans specifically have been seen as a negative and scary experience because of radiation and also because of the association that is directly made to a cancer diagnosis. Though, as mentioned in the chapter, this is not always the case and PET scans do have a variety of other functions. PET scans are readily available and common practice in Western, Educated, Industrialized, Rich, and Democratic societies such as Canada and the United States. In places such as Ontario, Canada they are even covered by local health insurance. However, there is a large disparity in other countries around the world which could be due to the fears surrounding radiation, but also the expense surrounding PET and/or PET-CT scanners. Lastly, sources have been provided in order to avoid misinformation.

The websites and documents mentioned are meant to guide patients and their families in order to ease the anxiety of having certain tests and procedures done. The one document specifically by the WHO is meant to show the specific dialogue between physicians and patients which should be had in order for both parties to feel comfortable.

The next chapter will cover where PET scan research is headed in the future. As hinted in previous chapters, the PET/MRI hybrid imaging technology may be one emerging discovery.

Where is PET Scan Research Headed in the Future?
By Razan Ahmed

Digital PET Scans

Medical devices typically use digital and analog technology combinations that transmit signals through cables or air on electromagnetic radiation. These signals are time-different in numbers, with encoded data (The promising future of digital pet scans, n.d).

PET scans show a radiotracer distribution inside the body of the sufferer with accumulations of possible cancer-cell indicators. The analog device uses traditional PET imaging devices to monitor scintillation bursts and transmit them to an optical screen. These light bursts occur as tracer released photons interfere with detector crystals (The promising future of digital pet scans, n.d). Clear pictures can imply that a patient is treated faster and the optimal results can be achieved. This is a guarantee with digital PET scans.

The adoption of digital PET scans at hospitals and other medical facilities will help preserve the lives of patients.

Hybrid PET/CT Scans

For some diseases such as cancer, coronary illness and certain cerebrum problems, the current and average imaging measure requires the securing of both positron emission tomography (PET) and computed tomography (CT) checks (Kjaer, 2014). PET scanners regularly measure metabolic movement, while CT scanners feature anatomical highlights. As these modalities are gained freely, there much of the time is an additional weight on the patient to need to make a trip to isolate imaging offices, and it requires more opportunity for the doctor to assess the scans (Kjaer, 2014).

Regularly a total assessment requires the doctor to see co-enrolled PET and CT intertwined pictures. The test in this methodology lies in precisely superimposing a patient's life structures across various modalities when pictures have been gained in independent sittings, with unstandardized boundaries, and by various technologists (Kjaer, 2014).

As of late, this issue has been demonstrated to be eased by the expanded accessibility of multimodality scanners equipped for procuring reciprocal imaging sets at a time (Kjaer, 2014). Clinically, the best illustration of multimodality imaging up to this point is in the utilization of PET/CT scanner cross breeds, which join the qualities of the two grounded imaging modalities to all the more precisely analyze, confine and screen illness. The danger to helpless superimposition of discrete imaging is moderated and, critically, the half and half PET/CT measure diminishes encumbrance on the patient by improving on the filtering planning measure (Kjaer, 2014).

Future of Brain PET

Brain PET has made its most prominent advances in measuring digestion and G-protein coupled receptors in different infection states and after treatment. Comparable, however lesser, achievement has been accomplished in the imaging of catalysts, e.g., monoamine oxidase An and B (MAO-An and - B), hexokinase, phosphodiesterase 4 and 10 (PDE4 and 10), unsaturated fat amide hydrolase (FAAH), and carriers, e.g., dopamine carrier (DAT), serotonin carrier (SERT/5-HTT), norepinephrine carrier (NET), vesicular monoamine carrier type 2 (VMAT2), and vesicular acetylcholine carrier (VAChT) (Jones & Rabiner, 2012). Be that as it may, it has been independently ineffective in the measurement of particle channels (other than GABA-A), or the intracellular second courier frameworks. The evaluation of kinase and phosphatase movement and insusceptible cycles stays in its early stages, as does the measurement of obsessive stores found in neurodegenerative illnesses (e.g., τ-protein and α-synuclein) (Jones & Rabiner, 2012). These objectives are of expanding significance in the examination of mind pathophysiology just as in the evaluation of novel medicines for CNS problems.

An area of extraordinary excitement in the research realm is the discovery of neurotransmitter changes in the living human cerebrum. While this has been grounded for the dopamine framework and has created important new information in understanding its job, expansion to different frameworks has been hampered by the absence of reasonable devices. Inside the previous 10 years, and specifically, in the last scarcely any, encouraging devices have been made accessible for imaging vacillations in 5-HT, endogenous narcotics, GABA, and glutamate (Jones & Rabiner, 2012). The blend of tissue pharmacokinetic information procured with PET, with pharmacodynamic information gained by other practical methods, for example, blood oxygen level-subordinate fMRI, FDG PET, and electroencephalography vows to give added esteem (Jones & Rabiner, 2012).

Advancement of Brain PET

PET information assortment, reproduction, show, and investigation are to some degree further developed than the applications. Notwithstanding, going ahead, there is still an extension for methodological upgrades. For instance, the makers of PET scanners could be urged to execute programming for tolerating applications so that reviews become less bulky and tedious (Jones & Rabiner, 2012). Expanding the accessibility of brain tomographs that carry out late specialized improvements would upgrade cerebrum pharmacokinetic gauges in microdose examinations. The foundation of unified photoshops at worldwide master places for particular quantitative examination of electronically moved dynamic PET information could likewise energize the more extensive advantageous utilization of this system (Jones & Rabiner, 2012).

The pharmaceutical business is in a situation to be a significant accomplice in this revelation and advancement measure, both as a client of mind PET to help drug improvement and as the holder of libraries of up-and-comer radioligands (Jones & Rabiner, 2012). An empowering improvement is an expanding acknowledgement inside Pharma and Biotech organizations that the advancement of novel radioligands addresses a precompetitive action, with data sharing inside industry and the scholarly community promising to propel the field by forestalling duplication of exertion (Jones & Rabiner, 2012). Multipartner agreeable drives are in progress, e.g., the European Innovative Medicine Initiative, which centers around the turn of events and approval of these devices. The responsibility of global scientists and physicists to take part in radioligand improvement is fundamental in finding and changing over promising applicant particles. Such responsibility should be conceivable, given the energizing potential for exploring the human mind. Yet, the way to prepare these researchers is improving the availability to fleeting positron-emanating radionuclides and the advancement of simpler tracer radiolabeling (Jones & Rabiner, 2012).

Alzheimer's Disease Recognition

One of the distinctive highlights of Alzheimer's illness is the development of harmful plaques of amyloid protein in the cerebrum. As of not long ago, the best way to identify this trademark was through posthumous tests on cerebrum tissue. Presently, because of amyloid PET imaging, specialists can recognize Alzheimer's mind plaques in living individuals. With this sort of clinical imaging, the individual gets an infusion of a synthetic tracer before they go through the sweep. The tracer goes to the mind and adheres to any amyloid plaques that may be available. These at that point appear on the sweep. In spite of the fact that there is no solution for Alzheimer's sickness, the capacity to analyze it all the more precisely in the beginning phases can assist specialists with recommending the right therapy and give patients and their families time to get ready for what's to come. It can likewise improve the choice of appropriate contender for Alzheimer's medication preliminaries.

Researchers at the Wisconsin Alzheimer's Disease Research Center and globally have discovered proof that tau PET scans are profoundly explicit to Alzheimer's infection. In an examination distributed in December 2019 reasoned that the presence of both amyloid and tau protein stores are related with psychological decay during the preclinical period of Alzheimer's (Research points to tau PET scans as the future of Alzheimer's disease diagnosis, 2020).

In future years, the explicitness of tau tracers to Alzheimer's illness could permit scientists to pinpoint the wellspring of a patient's intellectual decrease. Eminently, the first tau PET imaging tracer Tauvid was endorsed by the FDA to help in the conclusion of Alzheimer's infection, however suitable use models are yet to be set up (Research points to tau PET scans as the future of Alzheimer's disease diagnosis, 2020).

ImmunoPET In Cancer

PET is assuming an inexorably significant part in the finding, arranging, and checking reaction to therapy in an assortment of malignancies. Ongoing endeavors have zeroed in on ImmunoPET, which utilizes neutralizer based radiotracers, to picture tumors dependent on articulation of tumor-related antigens. It is proposed that the particularity managed by neutralizer focusing on ought to both improve tumor location and give phenotypic data identified with essential and metastatic injuries that will direct treatment choices. Advances in immunizer designing are giving the instruments to create counter acting agent based particles with pharmacokinetic properties upgraded for use as immunoPET radiotracers (Reddy, S., & Robinson, M. K, 2010). Combined with specialized advances in the plan of PET scanners, immunoPET holds guarantee to improve indicative imaging and to manage the utilization of focused treatments.

The extended utilization of monoclonal antibodies and focused on little atom inhibitors that are explicit for layer bound tumor-related antigens, proposes expected jobs for immunoPET in directing treatment choices (Reddy, S., & Robinson, M. K, 2010). ImmunoPET is appropriate to affirm antigen articulation in non-biopsied sores to empower determination of patients who are probably going to react to treatment, or to configuration educated elective treatment systems to improve patient reaction or potentially to stay away from superfluous treatment-instigated poison levels. ImmunoPET can be imagined to advise on the biodistribution and pharmacokinetics of new mAb treatments, both during preclinical and clinical turn of events, to expect likely poison levels. Moreover, immunoPET can be utilized to give dosimetry information in the setting of radioimmunotherapy (RAIT) for portion arranging (Reddy, S., & Robinson, M. K, 2010).

In the future, insusceptible based PET imaging could likewise empower specialists to screen patients over the long haul in light of different medicines. This would permit specialists to know whether a treatment is working, or on the off chance that it very well may be an ideal opportunity to think about potential next choices for the patient. Empowering simpler investigation of safe reaction to tumors over the long haul in the two individuals and creatures may help improve our comprehension of an assortment of resistant related cycles inside the setting of disease (Brodsky, 2020). This is remembered for its profundity portrayal of an assortment of insusceptible cell populaces, including B cells, administrative T cells, macrophages, regular executioner cells, and dendritic cells. These bits of knowledge could eventually uncover new components that are essential for fruitful enemies of malignant growth resistant reactions and give a reasoning to the advancement of cutting edge immunotherapies that focus on these cells and pathways (Brodsky, 2020).

Significant difficulties of new focused on treatment approaches include: 1) choice of patients that are probably going to react to therapy, 2) ID of the naturally dynamic fixation and proper dosing timetable, and 3) evaluation of the tumor reaction to treatment (Reddy, S., & Robinson, M. K, 2010). Future work should zero in on the reconciliation of immunoPET into the interaction of medication improvement and to address key inquiries in the preclinical and clinical assessment of novel focused on specialists with exceptional respect to the imaging of articulation and hindrance of medication targets, pharmacokinetics of new medications, and early evaluation of the tumor reaction to treatment. To satisfactorily evaluate clinical endpoints, markers that permit exact estimation of tumor focuses on an entire self endless supply of a useful specialist are required. Such specialists are required to give picture guided treatment that may permit end of insufficient medicines from the get-go over the span of treatment, and guide substitute more productive treatment methodologies that would be gainful to patients (Reddy, S., & Robinson, M. K, 2010).

Whole Body PET

Given the expanding and predicted more extensive utilization of entire body PET, it is expected that the presentation of new PET scanner innovation will address this developing difficulty by an expansion in affectability (Jones & Townsend, 2017). It is anticipated that this future innovation will be broadly received, bringing about new uses of PET in clinical exploration and medical care and along these lines understanding the maximum capacity of this imaging methodology. To accomplish this objective, advancement will be expected of physicists and architects empowered and animated by new applications including an entire body, frameworks science approach for misusing the uniqueness of PET-based sub-atomic imaging. Since entire body scanners currently conceal to 25 cm pivotally, the whole human mind can be imaged in a solitary bed position, blocking the requirement for a cerebrum just scanner. One exemption is the convenient PET scanner grew as of late to picture portable patients as they perform ordinary errands (Jones & Townsend, 2017).

There are an extensive number of current and possible future clinical applications for PET, and specifically requests for entire body imaging. In any case, in spite of all the advancement in the innovation portrayed above, there are as yet significant specialized restrictions to understanding this potential, which, thus, highlight the requirement for much more advances in PET instrumentation (Jones & Townsend, 2017). Effective affectability is lost at higher check rate because of the degree of irregular fortuitous events and indicator dead time. To limit these impacts, the occurrence timing window and consolidation of effective equal information assortment are approaches to decrease these wellsprings of loss of sensitivity. Given the characteristic affectability and particularity of PET, there is a case for underwriting further improvements to date by investigating the methods for distinguishing lower levels of pathophysiological measures than can be as of now accomplished (Jones & Townsend, 2017). This could be acknowledged either by improving spatial goal or utilizing active marks to uncover levels of existing illness even underneath the spatial goal of the scanner. Instances of the latter would be in the identification of the presence of micrometastases that are communicated as a different tracer motor part contained inside the dynamic segments of ordinary tissue. Such an investigation could likewise be applied to identification of low degrees of aggravation and disease (Jones & Townsend, 2017).

Preclinical Cardiovascular PET/MRI

There are a few zones where the use of consolidated PET/MRI may extraordinarily profit preclinical cardiovascular imaging soon. As shown by a few preclinical and clinical atherosclerosis examines, PET/MRI offers the particular advantage of permitting co-confinement of the PET sign with various plaque highlights, (for example, divider thickness, or arrangement) assessed by multi-parametric MRI, utilizing either unique MR weightings, or readouts, (for example, perfusion imaging). As effectively portrayed, this exceptional methodology permits the mix of various multi-methodology imaging highlights for top to bottom plaque phenotyping (Calcagno et al., 2020). Moreover, the high-goal and unrivaled delicate tissue difference of MR pictures, permit the utilization of fractional volume mistake adjustment procedures to PET pictures, to improve the meaning of vascular PET tracer take-up. Incomplete volume blunders comprise in the tainting of sign from adjoining picture voxels, prompting off base evaluation. These curios are especially pertinent to vascular imaging, either clinical or preclinical, because of the characteristically little size of the blood vessel divider (Calcagno et al., 2020). A few clinical investigations have utilized high-goal vascular MR information to improve the goal of PET images, and along these lines, the evaluation of vascular radiotracer take-up. In little creatures, halfway volume mistake adjustment procedures have been utilized, for instance, to improve the picture determined evaluation of blood tracer movement (the alleged blood vessel input work) in the bunny stomach aorta and in the mouse left ventricle,to help dynamic imaging contemplates (Calcagno et al., 2020).

Consolidated PET/MRI may likewise demonstrate major in improving quality and difference of PET pictures of the moving heart,for the more precise appraisal of vascular irritation or myocardial practicality after ischemia or for evaluation of the coronary arteries (Calcagno et al., 2020). Different than PET/CT, where cardiovascular and respiratory movement are commonly managed by gating obtained PET information based on outside ECG screens and respiratory howls, incorporated PET/MRI considers movement adjustment of the PET information utilizing the powerful procurement of MR pictures with high spatiotemporal goal during the heart and respiratory cycles. From these MR pictures, movement vector fields can be determined and applied to the entire arrangement of PET outflow information, to remake high sign to-clamor, movement rectified PET pictures (Calcagno et al., 2020). Moreover, this methodology permits assessing constriction maps during the cardiovascular and respiratory cycles, subsequently wiping out conceivable picture curios because of movement in the weakening guides themselves. At last, PET/MRI may be instrumental to the advancement of savvy, bimodal imaging tests for better portrayal of pathophysiological measures in the vasculature or different organs associated with cardiovascular illness (Calcagno et al., 2020).

References

What was Medical Imaging Like Before PET Scans?

Ambrose, E., Gould, T., & Uttley, D. (2006). Jamie Ambrose. BMJ, 332(7547), 977. doi: 10.1136/bmj.332.7547.977

Benson, C., & Doubilet, P. (2014). The History of Imaging in Obstetrics. Radiology, 273(2S), S92-S110. doi: 10.1148/radiol.14140238

Campbell, J. (2015). History and Development of MRI Scanning. Retrieved 4 May 2021, from http://large.stanford.edu/courses/2015/ph241/campbell1/

Chodos, A., & Ouellette, J. (2001). November 8, 1895: Roentgen's Discovery of X-Rays. Retrieved 3 May 2021, from https://www.aps.org/publications/apsnews/200111/history.cfm

Contrast Media. (2016). Retrieved 3 May 2021, from https://radiologykey.com/contrast-media-2/#:~:text=Radiographic%20contrast%20has%20been%20used,radiographic%20examinations%20of%20the%20chest.&text=Iodine%2Dbased%20contrast%20media%20have%20been%20used%20ever%20since.

Drozdovitch, V., Brill, A., Mettler, F., Beckner, W., Goldsmith, S., & Gross, M. et al. (2014). Nuclear Medicine Practices in the 1950s through the Mid-1970s and Occupational Radiation Doses to Technologists from Diagnostic Radioisotope Procedures. Health

Physics, 107(4), 300-310. doi: 10.1097/hp.0000000000000107

First Clinical X-ray in America Performed. Retrieved 5 May 2021, from https://250.dartmouth.edu/highlights/first-clinical-x-ray-america-performed

Freiherr, G. (2014). The Eclectic History of Medical Imaging. Retrieved 3 May 2021, from https://www.itnonline.com/article/eclectic-history-medical-imaging

Goel, A., & Bell, D. Cassette | Radiology Reference Article | Radiopaedia.org. Retrieved 3 May 2021, from https://radiopaedia.org/articles/cassette#:~:text=In%20the%20case%20of%20conventional,imaging%20plate%20in%20the%20light.

Half A Century In CT: How Computed Tomography Has Evolved — ISCT. (2016). Retrieved 3 May 2021, from https://www.isct.org/computed-tomography-blog/2017/2/10/half-a-century-in-ct-how-computed-tomography-has-evolved

Haus, A., & Cullinan, J. (1989). Screen film processing systems for medical radiography: a historical review. Radiographics, 9(6), 1203-1224. doi: 10.1148/radiographics.9.6.2685941

History of the CT Scan | Catalina Imaging. (2021). Retrieved 3 May 2021, from https://catalinaimaging.com/history-ct-scan/

History of Ultrasound – Overview of Sonography History and Discovery. Retrieved 3 May 2021, from https://www.ultrasoundschoolsinfo.com/history/

Hutchison, C. (2011). 1896 X-Ray Machine Shows Radiation Risks of Yore: 1,500 Times More Radiation Than Today. Retrieved 3 May 2021, from https://abcnews.go.com/Health/Wellness/century-ray-machine-shows-radiation-risks-yore/story?id=13140857#:~:text=As%20a%20result%2C%20experimenters%20using,hair%2C%20and%20 ultimately%2C%20cancer.

Intensifying screens. Problems and developments. (1955). Acta
Radiologica, 43, 66-80. doi: 10.3109/00016925509170732

Nadrljanski, M., & Bell, D. (2021). History of ultrasound in
medicine | Radiology Reference Article | Radiopaedia.org.
Retrieved 4 May 2021, from https://radiopaedia.org/articles/
history-of-ultrasound-in-medicine

National Academies Press. (2007). Advancing nuclear medicine
through innovation. Washington, D.C.

Newman, T. (2018). X-ray exposure: How safe are X-rays?.
Retrieved 3 May 2021, from https://www.medicalnewstoday.
com/articles/219970

Nuclear Medicine Market Value Projected To Reach US$ 12.8
Billion By 2027: Acumen Research And Consulting. (2021).
Retrieved 5 May 2021, from https://finance.yahoo.com/news/
nuclear-medicine-market-value-projected-141300261.html

PET Scan: Tests, Types, Procedure. (2021). Retrieved 3 May 2021,
from https://my.clevelandclinic.org/health/diagnostics/10123-
pet-scan#:~:text=Magnetic%20resonance%20imaging%20
(MRI)%20scans,than%20CT%20and%20MRI%20scans.

Riley, S. (2019). Anatomy professor uses 500-year-old da Vinci
drawings to guide cadaver dissection. Retrieved 4 May 2021, from
https://www.pbs.org/wgbh/nova/article/leonardo-da-vinci-
anatomy-dissection/

Rubin, A. (2021). History of Medical Imaging - A Brief Overview
- Health Beat. Retrieved 3 May 2021, from https://www.
flushinghospital.org/newsletter/history-of-medical-imaging-a-
brief-overview/#:~:text=The%20concept%20of%20medical%20
imaging,photosensitive%20plate%20placed%20behind%20it.

Sansare, K., Khanna, V., & Karjodkar, F. (2011). Early victims of X-rays: a tribute and current perception. Dentomaxillofacial Radiology, 40(2), 123-125. doi: 10.1259/dmfr/73488299

The History of Anatomy - from the beginnings to the 20th century. Retrieved 4 May 2021, from https://bodyworlds.com/about/history-of-anatomy/

The History of Ultrasound | BMUS. Retrieved 3 May 2021, from https://www.bmus.org/for-patients/history-of-ultrasound/#:~:text=The%20use%20of%20ultrasound%20in,published%20work%20on%20medical%20ultrasonics.

The Image Intensifier (II) | Radiology | SUNY Upstate Medical University. Retrieved 3 May 2021, from https://www.upstate.edu/radiology/education/rsna/fluoro/iisize.php

When and why was MRI invented. (2019). Retrieved 3 May 2021, from https://www.gehealthcare.com/article/when-and-why-was-mri-invented

Woo, J. (2015). History of Ultrasound in Obstetrics and Gynecology. Retrieved 3 May 2021, from https://www.ob-ultrasound.net/dussikbio.html

Who Invented or Discovered PET Scans?

Alavi A, Moghbel M, Alavi JB. Abass Alavi. A distinguished physician scientist and a pioneer in molecular imaging. Hell J Nucl Med. 2014 May-Aug;17(2):74-7. doi: 10.1967/s002449910130. Epub 2014 Jul 5. PMID: 24997075.

Brownell, G. L. (1999). A History of Positron Imaging. Retrieved from http://www.umich.edu/~ners580/ners-bioe_481/lectures/pdfs/Brownell1999_historyPET.pdf

Charlton M and Humberston JW. Positron Physics. Cambridge University Press, 2001, p. 6.

Dunnick, R. N. (2017). David E. Kuhl, MD. Retrieved from https://doi.org/10.1148/radiol.2017174024

G. L. Brownell, C. A. Burnham and D. A. Chesler, "High resolution tomograph using analog coding" in The Metabolism of the Human Brain Studied with Positron Emission Tomography, New York:Raven, pp. 13-19, 1985.

Høilund-Carlsen PF. Abass Alavi: A giant in Nuclear Medicine turns 80 and is still going strong! Hell J Nucl Med. 2018 Jan-Apr;21(1):85-87. doi: 10.1967/s002449910713. Epub 2018 Mar 20. PMID: 29550853.

Massachusetts Institute of Technology, Gordon L. Brownell, PROFESSOR emeritus, 86. (2008, November). Retrieved May 07, 2021, from https://news.mit.edu/2008/obit-brownell-1118#:~:text=Born%20in%20Duncan%2C%20Okla.%2C,PhD%20in%20physics%20from%20MIT.&text=Brownell%20established%20the%20Physics%20Research,at%20MGH%20until%20his%20death.

Medical radioisotope scintigraphy: Proceedings of a symposium on medical radioisotope scintigraphy held by the International Atomic Energy Agency in Salzburg, 6-15 august 1968 (Vol. 1). (1969). Vienna: International Atomic Energy Agency (IAEA).

M. R. Palmer and G. L. Brownell, "Annihilation density distribution calculations for medically important positron emitters," in IEEE Transactions on Medical Imaging, vol. 11, no. 3, pp. 373-378, Sept. 1992, doi: 10.1109/42.158941.

Patton D.D. (1980) Rectilinear Scanners. In: Nudelman S., Patton D.D. (eds) Nuclear Medicine, Ultrasonics,and Thermography. Springer, Boston, MA. https://doi.org/10.1007/978-1-4684-3671-6_3

Phelps, M. E., Hoffman, E. J., Mullani, N. A., Higgens, C. S., & Ter-Pogossian, M. M. (1976). Design Considerations for a Positron Emission Transaxial Tomograph (Vol. 1). Philadelphia, Pennsylvania: Department of Radiology, Hospital of the University of Pennsylvania.

Phelps, M., Hoffman, E., Mullani, N., & Ter-Pogossian, M. (1975, March 01). Application of annihilation coincidence detection to transaxial reconstruction tomography. Retrieved May 07, 2021, from https://jnm.snmjournals.org/content/16/3/210/tab-article-info

Portnow, L., Vaillancourt, D., & Okun, M. (2013). The history of cerebral PET scanning From physiology to cutting-edge technology. Neurology, 80(10), 952–956. https://doi.org/10.1212/WNL.0b013e318285c135

Rich, D. A. (1997). Special Contribution. In A Brief History of Positron Emission Tomography (1st ed., Vol. 25, pp. 4-11). New Haven, Conneticut: Yale University Medical School.

Rubin, A., & Rubin, A. (2017, April 06). History of medical imaging - a brief overview. Retrieved May 07, 2021, from https://www.flushinghospital.org/newsletter/history-of-medical-imaging-a-brief-overview/

Michel Ter-Pogossian (1925-96)". Science Museum, London. Archived from the original on 3 January 2020.

University of Pennsylvania health system: News and PERIODICALS: News releases. (n.d.). Retrieved May 06, 2021, from https://web.archive.org/web/20120426034842/

http://www.uphs.upenn.edu/news/News_Releases/june04/
HevesyAward.htm

Wackers, F.J.T., the History Corner. David E. Kuhl (1929–2017).
J. Nucl. Cardiol. 26, 1062–1063 (2019). https://doi-org.proxy.
library.brocku.ca/10.1007/s12350-018-1427-0

Wackers, Frans J. Th. (August 2018). "Michael M. Ter-Pogossian
(1925-1996)". Journal of Nuclear Cardiology. 25 (4): 1090–1091.
doi:10.1007/s12350-018-1313-9

Webb, S. (1990). Kuhl & Edwards PRODUCE CT Scan.
Retrieved May 06, 2021, from https://ethw.org/Kuhl_%26_
Edwards_Produce_CT_Scan

What are PET Scans?

American Heart Association. (2021). Positron emission
tomography. https://www.heart.org/en/health-topics/
heart-attack/diagnosing-a-heart-attack/positron-   emission-
tomography-pet

Brazier, Y. (2017, June 23). What is a PET scan and, are there
risks. https://www.medicalnewstoday.com/articles/154877#uses
Medical News Today. https://www.medicalnewstoday.com/
articles/154877#uses

Canadian Cancer Society. (2021). Positron emission tomography.
https://www.cancer.ca/en/cancer-information/diagnosis-and-
treatment/tests-and-   procedures/positron-emission-tomography-
pet-scan/?region=on#

Krans, B. (2018). What is a PET scan. Healthline. https://www.
healthline.com/health/pet-scan Mann, D. (2011, July 11). PET
scans may help with Alzheimer's diagnosis.

Web MD. https://www.webmd.com/alzheimers/ news/20110711/pet-scans-nay-help-with- alzheimers-diagnosis

Mayo Clinic. (2021). Positron emission tomography scan. https:// www.mayoclinic.org/tests- procedures/pet-scan/about/

Ruysscher, D.D., & Kirsch, C.M. (2010). PET scans for in radiotherapy planning lung cancer. Radiotherapy and Oncology, 96(3), 335-228. https://doi.org/https://doi.org/10.1016/j. radonc. 2010.07.002

University of Ottawa Heart Institute. (2021). PET viability imaging. https://www.ottawaheart.ca/test-procedure/pet-viability-imaging

Virtual Medical Centre. (2017). PET scan. https://www.myvmc. com/investigations/pet-scan- positron-emission-tomography

Weaver, C.H. (February 9, 2019). The role of PET scans in diagnosis and treatment of cancer.

Cancer Connect. https://news.cancerconnect.com/treatment-care/the-role-of-pet-scans-in-the- diagnosis-and-treatment-of-cancer?redir=1

What Science is Involved in PET Scans?

Alexander, S.J.I., Abrego, L., Ballesteros, M., Campos, E., Leon, D.D., & Fernandez, D. (2018).

Importance of chemistry in PET-CT studies. Biomarkers Journal. 4(4). doi: 10.21767/2472-1646.100042

Anosh. (2016, March 23). How Positron Emission Tomography works. Young Scientists Journal. https://ysjournal.com/how-positron-emission-tomography-works/.

Arpansa. (2019, May 14). Gamma radiation. Arpansa. https://www.arpansa.gov.au/understanding-radiation/what-is-radiation/ionising-radiation/gamma-radiation.

Berger A. (2003). How does it work? Positron emission tomography. BMJ (Clinical researched.), 326(7404), 1449. https://doi.org/10.1136/bmj.326.7404.1449

Cherry, S.R., Gambhir, S.S. (2001). Use of positron emission tomography in animal research.

ILAR Journal, 42(3), 219–232. doi: https://doi.org/10.1093/ilar.42.3.219

Gambhir, S. S. (2002). Molecular imaging of cancer with positron emission tomography. Nature

Reviews Cancer, 2(9), 683–693. https://doi.org/10.1038/nrc882

Harvard University. (2021). https://handbook.fas.harvard.edu/book/chemistry#:~:text=Chemistry%20is%20the%20science%20of,and%20variety%20of%20important%20applications.

Hietala, J. (2009). Ligand–receptor interactions as studied by PET: implications for drug development, Annals of Medicine, 31(6), 438-443. doi: 10.3109/07853899908998802

Jacobson, O., Kiesewetter, D.O., & Chen, X. (2015). Fluorine-18 radiochemistry, labeling strategies, and synthetic routes. Bioconjugate Chemistry, 26(1), 1-18. doi: 10.1021/bc500475e

Libretexts. (2020, September 22). Radiation in Biology and Medicine: Positron Emission

Tomography. Chemistry LibreTexts. https://chem.libretexts.org/Bookshelves/Physical_and Theoretical_Chemistry_Textbook_Maps/Supplemental_ Modules_(Physical_and_Theoretical_Chemistry)/Nuclear_ Chemistry/Applications_of_Nuclear_Chemistry/Radiation_in_ Biology_and_Medicine%3A_Positron_Emission_Tomography.

Naider, F. R., Brautigan, D. L., Grossman, J., Hugli, T. T., Lynch, R. G., Englander, S. W., & Redpath, J. L. (2011). Breakthroughs in bioscience. The FASEB Journal, 18(3). https://doi.org/10.1096/fasebj.18.3.421e

Schmitz, R.E., Alessio, A.M., & Paul E. Kinahan, P.E. (n.d.). The physics of PET/CT scanners. 1-16. https://depts.washington.edu/imreslab/education/ Physics%20of%20PET.pdf

Seeing More with PET Scans: Scientists Discover New Way to Label Chemical Compounds for Medical Imaging. News Center. (2017, July 27). https://newscenter.lbl.gov/2017/07/27/new-chemistry-pet-scans-medical-imaging/.

What Have We Learned About PET Scans in Recent Medical History?

Balink, H., Collins, J., Bruyn, G., & Gemmel, F. (2009). F-18 FDG PET/CT in the diagnosis of fever of unknown origin. Clinical Nuclear Medicine, 34(12), 862–868. https://doi.org/10.1097/ RLU.0b013e31811becfb1

Berthold, L. D., Steiner, D., Scholz, D., Alzen, G., & Zimmer, K. P. (2013). Imaging of chronic inflammatory bowel disease with 18F-FDG PET in children and adolescents. Klinische Pädiatrie, 225(4), 212–217. https://doi.org/10.1055/s-0033-1334878

Buchbender, C., Heusner, T. A., Lauenstein, T. C., Bockisch, A., & Antoch, G. (2012). Oncologic PET/MRI, part 2: Bone tumors, soft-tissue tumors, melanoma, and lymphoma. In Journal of Nuclear Medicine (Vol. 53, Issue 8, pp. 1244–1252). Society of Nuclear Medicine. https://doi.org/10.2967/jnumed.112.109306

Catana, C., Drzezga, A., Heiss, W. D., & Rosen, B. R. (2012). PET/MRI for neurologic applications. In Journal of Nuclear Medicine (Vol. 53, Issue 12, pp. 1916–1925). J Nucl Med. https://doi.org/10.2967/jnumed.112.105346

Cheng, C., Heiss, C., Dimitrakopoulou-Strauss, A., Govindarajan, P., Schlewitz, G., Pan, L., Schnettler, R., Weber, K., & Strauss, L. G. (2013). Evaluation of bone remodeling with (18)F-fluoride and correlation with the glucose metabolism measured by (18)F-FDG in lumbar spine with time in an experimental nude rat model with osteoporosis using dynamic PET-CT. American Journal of Nuclear Medicine and Molecular Imaging, 3(2), 118–128. http://www.ncbi.nlm.nih.gov/pubmed/23526138

Di Carli, M. F., Maddahi, J., Rokhsar, S., Schelbert, H. R., Bianco-Batlles, D., Brunken, R. C., & Fromm, B. (1998). Long-term survival of patients with coronary artery disease and left ventricular dysfunction: Implications for the role of myocardial viability assessment in management decisions. Journal of Thoracic and Cardiovascular Surgery, 116(6), 997–1004. https://doi.org/10.1016/S0022-5223(98)70052-2

Drzezga, A., Souvatzoglou, M., Eiber, M., Beer, A. J., Fürst, S., Martinez-Möller, A., Nekolla, S. G., Ziegler, S., Ganter, C., Rummeny, E. J., & Schwaiger, M. (2012). First clinical experience with integrated whole-body PET/MR: Comparison to PET/CT in patients with oncologic diagnoses. Journal of Nuclear Medicine, 53(6), 845–855. https://doi.org/10.2967/jnumed.111.098608

Eyuboglu, S., Angus, G., Patel, B. N., Pareek, A., Davidzon, G., Long, J., Dunnmon, J., & Lungren, M. P. (2021). Multi-task weak supervision enables anatomically-resolved abnormality detection in whole-body FDG-PET/CT. Nature Communications, 12(1), 1–15. https://doi.org/10.1038/s41467-021-22018-1

Garibotto, V., Heinzer, S., Vulliemoz, S., Guignard, R., Wissmeyer, M., Seeck, M., Lovblad, K. O., Zaidi, H., Ratib, O., & Vargas, M. I. (2013). Clinical applications of hybrid PET/MRI in neuroimaging. Clinical Nuclear Medicine, 38(1). https://doi.org/10.1097/RLU.0b013e3182638ea6

Grant, F. D., Fahey, F. H., Packard, A. B., Davis, R. T., Alavi, A., & Treves, S. T. (2008). Skeletal PET with 18F-fluoride: Applying new technology to an old tracer. Journal of Nuclear Medicine, 49(1), 68–78. https://doi.org/10.2967/jnumed.106.037200

Hany, T. F., Steinert, H. C., Goerres, G. W., Buck, A., & Von Schulthess, G. K. (2002). PET diagnostic accuracy: Improvement with in-line PET-CT system: Initial results. Radiology, 225(2), 575–581. https://doi.org/10.1148/radiol.2252011568

Hipp, S. J., Steffen-Smith, E. A., Patronas, N., Herscovitch, P., Solomon, J. M., Bent, R. S., Steinberg, S. M., & Warren, K. E. (2012). Molecular imaging of pediatric brain tumors: Comparison of tumor metabolism using 18F-FDG-PET and MRSI. Journal of Neuro-Oncology, 109(3), 521–527. https://doi.org/10.1007/s11060-012-0918-0

Hoehn, M., Wiedermann, D., Justicia, C., Ramos-cabrer, P., Kruttwig, K., Farr, T., & Himmelreich, U. (2007). Cell tracking using magnetic resonance imaging. In Journal of Physiology (Vol. 584, Issue 1, pp. 25–30). John Wiley & Sons, Ltd. https://doi.org/10.1113/jphysiol.2007.139451

Hoh, C. K., Hawkins, R. A., Dahlbom, M., Glaspy, J. A., Seeger, L. L., Choi, Y., Schiepers, C. W., Huang, S. C., Satyamurthy, N., Barrio, J. R., & Phelps, M. E. (1993). Whole body skeletal imaging with [18f]fluoride ion and pet. Journal of Computer Assisted Tomography, 17(1), 34–41. https://doi.org/10.1097/00004728-199301000-00005

Jacobs, A., Voges, J., Reszka, R., Lercher, M., Gossmann, A., Kracht, L., Kaestle, C., Wagner, R., Wienhard, K., & Heiss, W. D. (2001). Positron-emission tomography of vector-mediated gene expression in gene therapy for gliomas. Lancet, 358(9283), 727–729. https://doi.org/10.1016/S0140-6736(01)05904-9

Jamiel, A., Ebid, M., Ahmed, A. M., Ahmed, D., & Al-Mallah, M. H. (2017). The role of myocardial viability in contemporary cardiac practice. In Heart Failure Reviews (Vol. 22, Issue 4, pp. 401–413). Springer New York LLC. https://doi.org/10.1007/s10741-017-9626-3

Jasper, N., Däbritz, J., Frosch, M., Loeffler, M., Weckesser, M., & Foell, D. (2010). Diagnostic value of [18F]-FDG PET/CT in children with fever of unknown origin or unexplained signs of inflammation. European Journal of Nuclear Medicine and Molecular Imaging, 37(1), 136–145. https://doi.org/10.1007/s00259-009-1185-y

Khalaf, S., Chamsi-Pasha, M., & Al-Mallah, M. H. (2019). Assessment of myocardial viability by PET. Current Opinion in Cardiology, 34(5), 466–472. https://doi.org/10.1097/HCO.0000000000000652

Kluge, R., Kurch, L., Montravers, F., & Mauz-Körholz, C. (2013). FDG PET/CT in children and adolescents with lymphoma. Pediatric Radiology, 43(4), 406–417. https://doi.org/10.1007/s00247-012-2559-z

Lardinois, D., Weder, W., Hany, T. F., Kamel, E. M., Korom, S., Seifert, B., von Schulthess, G. K., & Steinert, H. C. (2003). Staging of Non–Small-Cell Lung Cancer with Integrated Positron-Emission Tomography and Computed Tomography. New England Journal of Medicine, 348(25), 2500–2507. https://doi.org/10.1056/nejmoa022136

London, K., Cross, S., Onikul, E., Dalla-Pozza, L., & Howman-Giles, R. (2011). 18F-FDG PET/CT in paediatric lymphoma: Comparison with conventional imaging. European Journal of Nuclear Medicine and Molecular Imaging, 38(2), 274–284. https://doi.org/10.1007/s00259-010-1619-6

Montravers, F., McNamara, D., Landman-Parker, J., Grahek, D., Kerrou, K., Younsi, N., Wioland, M., Leverger, G., & Talbot, J. (2002). [18F]FDG in childhood lymphoma: Clinical utility and impact on management. European Journal of Nuclear Medicine, 29(9), 1155–1165. https://doi.org/10.1007/s00259-002-0861-y

Purz, S., Mauz-Körholz, C., Körholz, D., Hasenclever, D., Krausse, A., Sorge, I., Ruschke, K., Stiefel, M., Amthauer, H., Schober, O., Kranert, W. T., Weber, W. A., Haberkorn, U., Hundsdörfer, P., Ehlert, K., Becker, M., Rössler, J., Kulozik, A. E., Sabri, O., & Kluge, R. (2011). [18F]fluorodeoxyglucose positron emission tomography for detection of bone marrow involvement in children and adolescents with Hodgkin's lymphoma. Journal of Clinical Oncology, 29(26), 3523–3528. https://doi.org/10.1200/JCO.2010.32.4996

Purz, S., Sabri, O., Viehweger, A., Barthel, H., Kluge, R., Sorge, I., & Hirsch, F. W. (2014). Potential pediatric applications of PET/MR. Journal of Nuclear Medicine, 55(6 SUPPL. 2), 32–39. https://doi.org/10.2967/jnumed.113.129304

Schirrmeister, H., Guhlmann, A., Elsner, K., Kotzerke, J., Glatting, G., Rentschler, M., Neumaier, B., Träger, H., Nüssle, K., & Reske, S. N. (1999). Sensitivity in Detecting Osseous Lesions

Depends on Anatomic Localization: Planar Bone Scintigraphy Versus 18F PET. Journal of Nuclear Medicine, 40(10).

Schlemmer, H. P. W., Pichler, B. J., Schmand, M., Burbar, Z., Michel, C., Ladebeck, R., Jattke, K., Townsend, D., Nahmias, C., Jacob, P. K., Heiss, W. D., & Claussen, C. D. (2008). Simultaneous MR/PET imaging of the human brain: Feasibility study. Radiology, 248(3), 1028–1035. https://doi.org/10.1148/radiol.2483071927

Singhal, T. (2012). Positron emission tomography applications in clinical neurology. Seminars in Neurology, 32(4), 421–431. https://doi.org/10.1055/s-0032-1331813

Townsend, D. W., & Beyer, T. (2002). A combined PET/CT scanner: The path to true image fusion. British Journal of Radiology, 75(SPEC. ISS.). https://doi.org/10.1259/bjr.75.suppl_9.750024

Von Schulthess, G. K., Steinert, H. C., & Hany, T. F. (2006). Integrated PET/CT: Current applications and future directions. Radiology, 238(2), 405–422. https://doi.org/10.1148/radiol.2382041977

Wu, L. M., Chen, F. Y., Jiang, X. X., Gu, H. Y., Yin, Y., & Xu, J. R. (2012). 18F-FDG PET, combined FDG-PET/CT and MRI for evaluation of bone marrow infiltration in staging of lymphoma: A systematic review and meta-analysis. In European Journal of Radiology (Vol. 81, Issue 2, pp. 303–311). Eur J Radiol. https://doi.org/10.1016/j.ejrad.2010.11.020

What are The Various Uses of PET Scans in Human and Animal Sciences?

Group name. (Year, Month Date). Title of page. Site name. URL
Krans, B. (2018). What Is a PET Scan? healthline. https://www.healthline.com/health/pet-scan

Weaver, C.H. (2019). The Role of PET Scans in the Diagnosis and Treatment of Cancer. Cancer Connect.https://news.cancerconnect.com/treatment-care/the-role-of-pet-scans-in-the-diagnosis-and-treatment-of-cancer?redir=1

Cancer.Net. (2020). Positron Emission Tomography and Computed Tomography (PET-CT) Scans.Cancer.Net.https://www.cancer.net/navigating-cancer-care/diagnosing-cancer/tests-and-procedures/positron-emission-tomography-and-computed-tomography-pet-ct-scans

Brazier, Y. (2017). What is a PET scan, and are there risks? Medical News Today.https://www.medicalnewstoday.com/articles/154877

Krans, B. (2018). Heart PET Scan. healthline. https://www.healthline.com/health/heart-pet-scan

Cardiovascular Institute of the South. (2018). Heart PET Scans: Who Needs Them and Why? Cardiovascular Institute of the South. https://www.cardio.com/blog/heart-pet-scans-who-needs-them-and-why

Virtual Medical Centre. (2017). Cardiac Pet Scan (Positron Emission Tomography). Virtual Medical Centre. https://www.myvmc.com/investigations/cardiac-pet-scan-positron-emission-tomography/#C4

Intermountain Healthcare. (2018). Positron Emission

Tomography Scan. Intermountain Healthcare.https://
intermountainhealthcare.org/services/pediatrics/services/
imaging-services/positron-emission-tomography-scan/

Gunville, L. (2018). WCVM to operate Canada's first PET-CT
unit for animals. University of Saskatchewan.https://news.usask.
ca/articles/colleges/2018/wcvm-to-operate-canadas-first-pet-ct-
unit-for-animals.php

Hill, A. (2019). 'On the precipice of a really big adventure;' U
of S becomes the first Canadian facility to have a PET-CT scan
for animal use. The Star Phoenix.https://thestarphoenix.com/
news/local-news/on-the-precipice-of-a-really-big-adventure-u-
of-s-becomes-the-first-canadian-facility-to-have-a-pet-ct-scan-for-
animal-use

Radiology.info. (2019). Positron Emission Tomography
- Computed Tomography (PET/CT). Radiology.
info.https://www.radiologyinfo.org/en/info/
pet#97f37e2b495d4ebf9e04656d8062e557

Stanford Health Care. (2020). What to Expect during a PET Scan.
Stanford Health Care.https://stanfordhealthcare.org/medical-
tests/p/pet-scan/what-to-expect.html

Better Health. (2019). PET scan. Better Health.https://www.
betterhealth.vic.gov.au/health/conditionsandtreatments/pet-
scan#when-the-pet-scan-is-used

Independent Imaging. (2021). PET Scan vs. CT Scan vs. MRI.
Independent Imaging.https://www.independentimaging.com/pet-
scan-vs-ct-scan-vs-mri/

Medical Imaging of Fredericksburg. (2019.) CT scan vs. a PET
scan. Medical Imaging of Fredericksburg.https://mifimaging.
com/2017/09/13/ct-scan-vs-a-pet-scan/

AICA Orthopedics. (2019). What's the Difference Between CT, MRI and PET Scans? AICA Orthopedics.https://aica.com/whats-the-difference-between-ct-mri-and-pet-scans/

Stanford Health Care. (2020). Risks of Magnetic Resonance Imaging (MRI). Stanford Health Care.https://stanfordhealthcare.org/medical-tests/m/mri/risk-factors.html

What are Opposing or Alternative Imaging Technologies to PET Scans?

Spine, M. B. &. (n.d.). SPECT scan. mayfieldclinic.com. https://mayfieldclinic.com/pe-spect.htm#:~:text=SPECT%20is%20a%20nuclear%20imaging,is%20injected%20into%20your%20bloodstream.

Medical information and health advice you can trust. (n.d.). https://www.healthline.com/.

Mayo Foundation for Medical Education and Research. (2019, August 3). MRI. Mayo Clinic. https://www.mayoclinic.org/tests-procedures/mri/about/pac-20384768.

Glover, G. H. (2011, April). Overview of functional magnetic resonance imaging. Neurosurgery clinics of North America. https://www.ncbi.nlm.nih.gov/pmc/articles/PMC3073717/.

What Misinformation or Conspiracy Theories Exist Regarding PET Scans?

Alimov, R. (2020, April 23). Chernobyl still burns. Greenpeace. https://www.greenpeace.org/international/story/30198/chernobyl-still-burns-forest-fires-u   kraine-nuclear-radiation/.

Berger, A. (2003). How does it work? Positron emission tomography. BMJ, 326 (7404), 1449. https://doi.org/10.1136/bmj.326.7404.1449

docpanel. (2020, May 18). Debunking cancer myths: The truth about 9 cancer imaging misconceptions. https://www.docpanel.com/blog/post/debunking-cancer-myths-truth-about-8-cancer-imaging-misconceptions.

Mayo Foundation for Medical Education and Research. (2020, August 25). Positron emission tomography scan. https://www.mayoclinic.org/tests-procedures/pet-scan/about/pac-20385078.

Portnow, L., Vaillancourt, D., and Okun, M. (2013). The history of cerebral PET scanning: From physiology to cutting-edge technology. Neurology, 80 (10), 952-956. https://doi.org/10.1212/WNL.0b013e318285c135.

Rohren, E. (2014). PET scanning: Worth the cost in cancer? Not only worth the cost, but sometimes a cost-cutter! Oncology, 28 (5), 390-392. Retrieved May 3, 2020, from https://www.cancernetwork.com/view/pet-scanning-worth-cost-cancer-not-only-worth-co   st-sometimes-cost-cutter

Safaie, E., Matthews, R., and Bergamaschi, R. (2015). PET scan findings can be false positive. Techniques in Coloproctology, 19 (1), 329-330. https://doi.org/10.1007/s10151-015-1308-3.

https://www.ncbi.nlm.nih.gov/pmc/articles/PMC3653214/
https://www.mayoclinic.org/tests-procedures/pet-scan/about/pac-20385078

https://www.docpanel.com/blog/post/debunking-cancer-myths-truth-about-8-cancer-imaging-misconceptions

https://www.greenpeace.org/international/story/30198/chernobyl-still-burns-forest-fires-ukraine-nuclear-radiation/

https://www.cancernetwork.com/view/pet-scanning-worth-cost-cancer-not-only-worth-cost-sometimes-cost-cutter

https://link.springer.com/article/10.1007/s10151-015-1308-3

How are PET Scans Talked About in Commonly and in Popular Culture?

The Mayo Clinic
  (https://www.mayoclinic.org/tests-procedures)

Radiologyinfo.org
  (https://www.radiologyinfo.org/en/test-treatment)

WHO: Ionizing Radiation Chapter 3- The Risk- Benefit dialogue
(https://www.who.int/ionizing_radiation/pub_meet/chapter3.
pdf?ua=1)

Additional links (not discussed):
a. Major hospitals in the United States Radiology Programs
1. Massachusetts General Hospital   (https://www.massgeneral.
org/imaging/programs-and-services/)

2. Johns Hopkins
  (https://www.hopkinsmedicine.org/imaging/)
b. Major hospitals in Canada Radiology Programs

3. Mount Sinai
(https://www.mountsinai.on.ca/care/me/medical-imaging)

4. Sunnybrook
(https://sunnybrook.ca/content/?page=dept-medimg-home)
c. International

5. St. Thomas Elgin General Hospital (https://www.stegh.on.ca/hospital-services/diagnostic-imaging/) London,UK

6. The Royal Melbourne Hospital (https://www.thermh.org.au/health-professionals/clinical-services/imaging/radiology) - Melbourne, Australia

Alex, J. I., S, er, Abrego, L., Ballesteros, M., Campos, E., Leon, D. D., ... Ez. (2018, February 27). Importance of Chemistry in Pet-Ct Studies. Biomarkers Journal. https://biomarkers.imedpub.com/importance-of-chemistry-in-petct-studies.php?aid=21953.

ACR, R. S. N. A. and. (2019, January 23). Magnetic Resonance, Functional (fMRI) - Brain. Radiologyinfo.org. https://www.radiologyinfo.org/en/info/fmribrain.

Ball, P. (2003, October 17). Lasers may make PET scans cheaper. Nature News. https://www.nature.com/news/2003/031013/full/news031013-11.html#:~:text=Radioactive%20materials%20for%20medical%20imaging%20produced%20at%20lower%20cost.&text=PET%20scanning%20could%20become%20cheaper,imaging%20technique%20cumbersome%20and%20expensive.

Positron Emission Tomography in Canada 2015. CADTH.ca. (2015, September 29). https://www.cadth.ca/positron-emission-tomography-canada-2015.

Carr, D. S. M. (2015). Radiation dosimetry: mSv & mGy. https://www.mun.ca/biology/scarr/Radiation_definitions.html

Chang, J. M., Lee, H. J., Goo, J. M., Lee, H. Y., Lee, J. J., Chung, J. K., & Im, J. G. (2006). False positive and false negative FDG-PET scans in various thoracic diseases. Korean journal of radiology, 7(1), 57–69. https://doi.org/10.3348/kjr.2006.7.1.57

Fernandez-Friera, L., García-Álvarez, A., & Ibáñez, B. (2013).

Imagining the future of   diagnostic imaging. Revista espanola de cardiologia, 66 2, 134-43 .

Gallach, Miguel et al. "Addressing Global Inequities in Positron Emission Tomography-Computed Tomography (PET-CT) for Cancer Management: A Statistical Model to Guide Strategic Planning." Medical science monitor : international medical journal of experimental and clinical research vol. 26 e926544. 27 Aug. 2020, doi:10.12659/MSM.926544

Goodman, E. (2019, April 4). Preparing for Your PET-CT Scan. Cancer.Net. https://www.cancer.net/blog/2019-04/preparing-your-pet-ct-scan#:~:text=Avoid%20exercising%2024%20 hours%20before,6%20hours%20before%20the%20scan.

Radiology Information System Market Size: Industry Report 2019-2026. Radiology Information System Market Size Industry Report 2019-2026. (n.d.). https://www.grandviewresearch.com/industry-analysis/radiology-information-system-ris-market.

Publishing, H. H. (2010, October). Radiation risk from medical imaging. Harvard Health. https://www.health.harvard.edu/cancer/radiation-risk-from-medical-imaging.

IAEA. (2014, November 21). Radiation in Everyday Life. IAEA. https://www.iaea.org/Publications/Factsheets/English/radlife.

Positron Emission Tomography (PET). Johns Hopkins Medicine. (n.d.). https://www.hopkinsmedicine.org/health/treatment-tests-and-therapies/positron-emission-tomography-pet#:~:text=PET%20works%20by%20using%20a,organ%20 or%20tissue%20being%20examined.&text=Gamma%20 rays%20are%20created%20during,then%20detects%20the%20 gamma%20rays.

Libretexts. (2020, September 22). Radiation in Biology and Medicine: Positron Emission

Tomography. Chemistry LibreTexts.

LSU Health Sciences. (2012, July). LSUSD Top Stories. https://
www.lsusd.lsuhsc.edu/News/KellsMuseum.html.

Mayo Foundation for Medical Education and Research. (2020,
February 28). CT scan. Mayo Clinic. https://www.mayoclinic.
org/tests-procedures/ct-scan/about/pac-20393675.

Mayo Foundation for Medical Education and Research. (2020,
August 25). Positron emission tomography scan. Mayo Clinic.
https://www.mayoclinic.org/tests-procedures/pet-scan/about/
pac-20385078.

Radiation Sickness. NORD (National Organization for Rare
Disorders). (2015, July 21). https://rarediseases.org/rare-diseases/
radiation-sickness/.

Orenstein. (2015, May). Three Scan Rule: Researchers Question
PET/CT Follow-Up Rule. https://www.radiologytoday.net/
archive/rt0515p24.shtml.

Poslusny, C. (2018, July 31). How much should your PET scan
cost? New Choice Health Blog. https://www.newchoicehealth.
com/pet-scan/cost#:~:text=The%20average%20PET%20
scan%20cost,or%20an%20outpatient%20surgery%20center.

Safaie, E., Matthews, R. & Bergamaschi, R. PET scan findings
can be false positive. Tech Coloproctol 19, 329–330 (2015).
https://doi.org/10.1007/s10151-015-1308-3

Sansare, K., Khanna, V., & Karjodkar, F. (2011). Early victims
of X-rays: a tribute and current perception. Dento maxillo
facial radiology, 40(2), 123–125. https://doi.org/10.1259/
dmfr/73488299

Simon, S. (2020, January 8). Facts & Figures 2020 Reports Largest One-year Drop in Cancer Mortality. American Cancer Society. https://www.cancer.org/latest-news/facts-and-figures-2020. html#:~:text=Facts%20%26%20Figures%202020%20 Reports%20Largest%20One%2Dyear%20Drop%20in%20 Cancer%20Mortality,-Written%20By%3AStacy&text=The%20 death%20rate%20from%20cancer,from%20the%20 American%20Cancer%20Society.

Snyder, C. (2014, December 29). Radiation in Pop culture. Online Radiology Technician Schools. https:// onlineradiologytechnicianschools.com/2014/radiation-in-pop-culture/.

Department of Labor logo UNITED STATES DEPARTMENT OF LABOR. Ionizing Radiation - Control and Prevention | Occupational Safety and Health Administration. (2014). https:// www.osha.gov/ionizing-radiation/control-prevention.

Functional MRI (fMRI). The University of Edinburgh. (2019, January 23). https://www.ed.ac.uk/clinical-sciences/edinburgh-imaging/research/themes-and-topics/medical-physics/imaging-techniques/functional-mri.

Usage Charges. Magnetic Resonance Research Center. (2020, July 1). https://medicine.yale.edu/mrrc/users/ charges/#:~:text=%24555%20per%20hour.,additional%20 30%20min%20increment%20options)

Where is PET Scan Research
Headed in the Future?

Brodsky, A. N. (2020, May 20). Immuno-Guided PET Scans May Improve Cancer Patient Treatment Decisions. Cancer Research Institute. https://www.cancerresearch.org/blog/may-2020/immuno-guided-pet-scans-cancer-patient-treatment.

Calcagno, C., Pérez-Medina, C., Mulder, W. J. M., & Fayad, Z. A. (2020, April 2). Digital Object Identifier System. https://doi.org/10.1161/ATVBAHA.119.313629.

Jones, T., Rabiner, E. A., & PET Research Advisory Company (2012). The development, past achievements, and future directions of brain PET. Journal of cerebral blood flow and metabolism : official journal of the International Society of Cerebral Blood Flow and Metabolism, 32(7), 1426–1454. https://doi.org/10.1038/jcbfm.2012.20

Jones, T., & Townsend, D. W. (2017, March 31). History and future technical innovation in positron emission tomography. Journal of Medical Imaging. https://www.spiedigitallibrary.org/redirect/medicalimaging/article?doi=10.1117%2F1.JMI.4.1.011013.

Kjaer A. (2014). Hybrid imaging with PET/CT and PET/MR. Cancer Imaging, 14(Suppl 1), O32. https://doi.org/10.1186/1470-7330-14-S1-O32

Reddy, S., & Robinson, M. K. (2010). Immuno-positron emission tomography in cancer models. Seminars in nuclear medicine, 40(3), 182–189. https://doi.org/10.1053/j.semnuclmed.2009.12.004

Research points to tau PET scans as the future of Alzheimer's disease diagnosis . Alzheimer's Disease Research Center. (2020, July 7). https://www.adrc.wisc.edu/news/research-points-tau-pet-scans-future-alzheimers-disease-diagnosis.

The promising future of digital pet scans. (n.d.). https://www.gehealthcare.com/article/the-promising-future-of-digital-pet-scans.